Garnet

Butterworths Gem Books
Edited by Peter G. Read

Beryl
John Sinkankas

Garnet
John D. Rouse

Pearls, natural, cultured and imitation
Alexander E. Farn

Quartz
Michael O'Donoghue

In preparation

Jet and amber
H. Muller and H. Franquet

Opals
P. J. Darragh

Topaz
D. B. Hoover

Butterworths Gem Books

Garnet

John D. Rouse

Butterworths
London Boston Durban Singapore Sydney Toronto Wellington

First published 1986

British Library Cataloguing in Publication Data

Rouse, John D.
Garnet.–(Butterworths gem books)
1. Garnet
I. Title
553.8′7 QE391.G37

ISBN 0-408-01534-9

Library of Congress Cataloging in Publication Data

Rouse, John D. (John David)
Garnet.
(Butterworths gem books)
Includes bibliographies and index.
1. Garnet. I. Title. II. Series.
QE391.G37R63 1986 553.8′7 85-17143

ISBN 0-408-01534-9

Photoset by Butterworths Litho Preparation Department
Printed and bound in Great Britain by The Garden City Press Ltd, Letchworth, Herts

Preface

Only a few books on garnet gemstones have been published up to the present time. The advantage of writing a book in a new area is very apparent – it is an entirely new subject for the reader. However, there are disadvantages from the standpoint of the author when a new field is undertaken, as established works provide starting places and a framework for research.

Consequently, this book is to be viewed, not as a finale in a massive undertaking, but rather as a pause to reflect on several years of research and absorption with garnets from the perspective of a cutter, dealer and gem enthusiast. The main purpose of this work is not to answer all the questions one might have about garnet, but rather to stimulate the readers to explore deeper into this very fascinating subject. It is a well known adage that the more one delves into a subject, the more one realizes how little one really knows about it.

Certainly there is an excitement derived in viewing gems, and it is hoped that this thrill is not mitigated by investigating the historical background and the scientific complexity of the garnet gemstones. On the other hand, as a result of this book, it is hoped that the reader will be inspired to collect, to study, to appreciate, and to wear the many garnet gemstones available in today's marketplace.

The author is indebted to many colleagues and scholars in various parts of the world for assistance in this project. Special thanks are owed to John Sinkankas, who made his extensive files on garnet available to the author. Also, appreciation is expressed to Roy Huddlestone for allowing me to work in his London offices and giving much of his time for this endeavour. Peter Read is to be thanked for his technical editing assistance, while Michael O'Donoghue of the British Museum libraries and Alan Jobbins of the Geological Museum library gave much help in explaining the resources of these very large library systems, and rendering research advice. Gratitude is also expressed to the professional staff of the Asian Institute of Gemological Sciences, who provided much time and work to this endeavour.

Special thanks are given to Richard Hughes for technical advice, Wimon Manorotkul for the many inclusion transparencies, and my secretary Pilaiphan for typing much of the manuscript. Mr W.K. Ho and Henry Ho are also to be acknowledged for their encouragement and assistance.

Lastly, but not the least important, I owe much to my wife Gloria, who proved of immense value in proof reading most of the text, following me to three continents and into many libraries as well as being an author's widow without complaining. To her, this work is dedicated.

Any errors and misunderstanding of facts, however, are my own responsibility.

John D. Rouse

Contents

PROFILE: SUMMARY OF CONSTANTS AND CHARACTERISTICS

RI (range)	*Dispersion*	*SG (range)*	*Hardness*
Almandine			
1.780–1.820	0.027	3.95–4.30	7½
Andradite			
1.880–1.888	0.057	3.77–3.88	6½
Grossular			
1.730–1.760	0.028	3.40–3.78	7
Pyrope			
1.730–1.750	0.022	3.65–3.80	7¼
Pyralspite series (pyrope-almandine and pyrope-spessartite)			
1.750–1.780	–	3.80–3.95	7–7¼
Spessartite			
1.790–1.810	0.027	4.12–4.20	7¼
Uvarovite			
1.87	–	3.77	7½

Species	*Chemical composition*	*Colour*
Almandine	$Fe_3Al_2(SiO_4)_3$	Orangy-red, red, violet-red
Andradite	$Ca_3Fe_2(SiO_4)_3$	Green (demantoid), greenish-yellow to yellow (topazolite)
Grossular	$Ca_3Al_2(SiO_4)_3$	Pink, orange-red, brownish-red, orange, orange-brown (hessanite) yellow, green, colourless
Pyrope	$Mg_3Al_2(SiO_4)_3$	Dark tones of red, orange-red, violet-red
Pyralspite (composition is a mixture of pyrope/almandine or pyrope/spessartite)		Violet-red (rhodolite, orange to orange-red (malaya), medium to dark orange to orange-red (pyrandine)
Spessartite	$Mn_3Al_2(SiO_4)_3$	Orange, orange-red, brownish-red, yellow-orange
Uvarovite	$Ca_3Cr_2(SiO_4)_3$	Emerald green, yellow-green

Crystal system Cubic

Habit Dodecahedron and trapezahedron crystals

Fluorescence Hydrogrossular variety of grossular fluoresces orange under X-rays

Absorption spectra See Appendix 2, page 125
Uvarovite, no characteristic spectrum

Principal occurrences Almandine; Austria, Brazil, USA, India, Sri Lanka
Andradite; USSR (demantoid), Switzerland and Italy (topazolite)
Grossular; Canada, South Africa (hydrogrossular), Mexico and Sri Lanka (especially hessonite)
Pyrope; Australia, Czechoslovakia, South Africa
Spessartite; Brazil, Burma, Malagasy Republic, Sri Lanka, USA
Uvarovite; Finland, Poland, USA, USSR

Chapter 1

The history of garnet

Students of gemstone history, analyzing ancient texts and museum artifacts for gemological clues in order to interpret often complex and contradictory evidence, may appear to be like detectives on a long trail of a criminal who committed only a trivial misdemeanor crime. The question must be asked, 'why bother'?

In the case of gemstones (and garnet gemstones in particular), without an understanding of their history, one is deprived of the struggle and effort behind landmark scientific achievements; of the knowledge concerning early concepts of aesthetic appreciation, often as modern as our own; of information on important mining and trade centres; of insights into ancient marketing conditions; and of the fascinating lore and romance usually associated with gems in ancient times. Historical studies are vital to a modern, comprehensive knowledge of gemstones.

The scattered pieces of the puzzle, confusing as they sometimes are, nevertheless tell a story. How well that story can be told depends in part, on the quantity and quality of the historical evidence and in part, on its interpretation. This chapter on the history of garnets is intended only as an inducement for further studies in a very intriguing and fascinating aspect of gemmology. There is much more material to be analyzed than can possibly be incorporated in the space allowed in this book. For students desiring to pursue further studies in this field, it is imperative to begin with the historical sources available to the modern researcher.

Historical evidence

The raw data are conveniently classified into three separate categories:

primary sources;
secondary sources;
archaeological artifacts.

The first two are literary accounts and the third covers specimens found in private collections, in the antique marketplace, or in various worldwide museums.

Primary sources

Among the primary sources are ancient writers whose works have survived. They include such authors as Theophrastus, Strabo, Pliny, Solinus, Damigeron, Epiphanius in the ancient period; Marbode and Albertus Magnus in the Medieval period; and Camillus Leonardus, Georgius Agricola, Boetius de Boot in the later period, to name but a few. Some of these writers exist in English translations, but many are still in the original Latin. Some knowledge of ancient Greek is helpful particularly for Theophrastus, in order to interpret fine points and conduct word studies. But there are two excellent translations available in the works of Hill (1774) and Eichholz (1965), with extensive notes. Hill's edition is interesting, because it reflects the knowledge on the red gemstones common in his day, thereby providing two primary sources – one for the text and one for the notes.

Pliny is available in the Loeb series, translated by Eichholz in Volume 10 (1962), corresponding to Pliny's books 36–37, the last two of the series (*Naturalis Historia*). Ball's translation of book 37 (1950) provides an alternate work, and, unlike the Loeb edition, incorporates extensive notes as well as a thorough discussion of many related topics.

Other Roman writers, especially Solinus, Damigeron and Epiphanius add little to the discussion of red gems by Pliny, but are useful in other respects. Solinus and Damigeron are available in the original Latin editions; Epiphanius can be found in English.

Of the medieval writers, Marbode, an eleventh century author, exists in an English translation found in one of King's works (King, 1866), and represents an important source of the period. Dorothy Wyckoff translated Albertus Magnus' 13th century work into English (1967), and Agricola's important work *de natura fossilium* (1546) was translated into English in 1955 and in German in 1958. Boetius de Boodt's landmark work (1609) has not yet been translated into English; it is, however, available in a French edition (1644).

Boyle's work (1672), although being of immense value in establishing mineralogy as a science, presents little information on red gemstones. The many scientific studies of the 18th century lay the foundation for mineralogy, and by the end of that century red gemstones are finally separated into the modern species, published by Haüy in 1801 and other works thereafter.

A recurring theme of magic and medicinal uses follows the gemstones through the ages. Yet Theophrastus and Pliny were not involved to any serious degree in either topic. Theophrastus was merely discussing gemstones from a philosophical point of view, following Aristotle; Pliny was a sceptic who relied on facts, either reported by personal observation, or through the writings of others. Although both writers did report some examples of both magic and medicine regarding gemstones, neither one wrote in the same manner as did the Medieval authors. The latter works were filled with many stories of medicinal remedies and magic associated with gemstones. These themes dominated the literature from about 148 AD up to the 18th century.

Secondary sources

In addition to many primary sources available to the student of gemstone history, a large number of secondary sources must be investigated. Historical works are necessary to understand the periods of the writers. Geological and geographical studies are important for investigating the existence, scope and location of the ancient mines. Articles in *Economic Geology*, the *American Mineralogist*, Chapter IX of Ball's translation of Pliny (1950), and numerous other periodicals include important information.

In determining the demand for gemstones in the ancient and Medieval world, one must explore gemstone use in jewellery. There are many basic texts available in this field. Greek and Roman gems are well illustrated and discussed in G.M.A. Richter's classic work, *Catalogue of the Engraved Gems, Greek, Etruscan and Roman* (1956), Adolph Furtwangler's monumental work, *Die antiken Gemmen* (1900) in three volumes, John Boardman's work, *Greek Gems and Finger Rings* (1970), George F. Kunz' work, *Rings for the Finger* (1917), Joan Evan's, *Magical Jewels* (1922), many of C.W. King's works, but especially *Antique Gems: their origin uses and value* (1866), and his *Antique Gems and Rings* (1872). Also R.A. Higgin's work, *Greek and Roman Jewellery* (1981, 2nd edition), and J.M. Ogden's *Jewellery of the Ancient World: The Materials and Techniques* (1982) should be included. Ball's translation of Pliny (1950) also contains several well-documented chapters on Roman jewellery, Roman jewellers and lapidaries. For trade developments from age-old gem sources of the East to the West, E.H. Warmington's work, *The Commerce Between the Roman Empire and India* (1928) is an authoritative and often-quoted source.

Summaries of ancient to eighteenth century lapidary writers in general can be found in Evan's book, *Magical Jewels* (1922), John Sinkankas' book on *Beryl* (1981), and Dorothy Wyckoff's translation of *Albertus Magnus* (1967).

Artifacts

Artifacts of ancient or medieval gemstones or jewellery are found in many museums and are published in catalogues. The British Museum's collection is brought together by Marshall, *Catalogue of the Jewellery, Greek Etruscan and Roman in the Department of Antiquities, British Museum* (1911), the *Catalogue of the Finger Rings, Greek Etruscan and Roman in the Department of Antiquities, British Museum* (1970), by the same author and Walters, *Catalogue of Engraved Gems, Cameos, Greek, Etruscan and Roman in the Britsh Museum* (1926). Shirley Bury's work, *Jewellery Gallery Summary Catalogue* (1983) is a descriptive catalogue of the extensive jewellery collection in the Victoria and Albert Museum. The Ashmolean collection is published as *Finger Rings from Ancient Egypt to the Present Day*, by Gerald Taylor and Diana Scarisbrick (1978). Many other museum collections, some quite major, can also supplement the above list.

The difficulties with historical gemmological research are found in the many specialties that must be investigated. Studies in geology, geography, history, economic history, ancient gemstones and jewellery, and Greek and Latin literature emphasize the point. The art forms on signets are quite appropriate studies for the art historian but the gemmologist-historian must confine his investigations to the gems themselves – their colour, inclusions, physical and optical properties, values, and their sources. These latter subjects provide the scope for the following analyses on ancient and later writers on the subject of the red gemstones, primarily garnet.

The fiery gemstones

Unfortunately, the modern understanding and appreciation of garnet gems is suffering an all-time low, due to a massive saturation of inexpensive Victorian-type jewellery that flooded the marketplace from about the mid 1850s until the early 1900s in both America and England. Fuelled by a new wealth among the upper industrial class of workers in both countries and a new consumer awareness of fine material goods, the movement spread rapidly. The focal point for such a vast market centered in Bohemia, an ancient source of pyrope garnets.

Even though the market was rapidly saturated, the legacy left in the memories of the people still lingers on. Garnet, to most people today, is an unattractive, very dark red stone found in their grandmother's rings. With the discovery of other colours and many other hues and tones available in this gemstone, the esteem is, however, beginning to recover.

In ancient times garnet was in a special category of gems, known as the 'fiery gemstones'. Ancient peoples were intrigued by the fire-like qualities of the garnets, as well as the variations in colour. With the curiosity of children, they speculated on their origin and studied them under lamplight, firelight and sunlight.

Of course their interest was coloured by strong beliefs in magic, the supernatural and seemingly inexplicable natural forces. A great mystique and reverence was afforded to the sun, moon and the awesome power of the lightning bolt. In the garnet they saw a reflection of that power which was seemingly captured within this red gem. Although to some the gem was a reflection of this power and glory to others it became an amulet to ward off pestilence or illness; and to still others it was a pretty ornament to be worn as a ring or necklace. To many Romans, in particular, it became a carved intaglio seal-ring for signing official documents.

That garnets were utilized in pre-history is well-attested. Bohemian garnet necklaces dating from the Bronze Age have been found in graves near the mining source (Ball, 1950, p. 53). Furthermore, necklaces containing garnet are frequently found among the ornaments adorning the oldest Egyptian mummies (Farrington, 1903, p. 128). In the Western World, during the historical periods, garnet gems and jewellery have been found in virtually every century for the past 2500 years. Although their popularity fluctuated, they were not always the most highly favoured gemstones. However, the existence of numerous examples of garnet gems and jewellery in archaeological sites and world-wide museums demonstrates clearly that garnet was widely utilized.

One of the first gemmological descriptions given in the literature was that recorded in the Bible, when gemstones were listed for the adornment of Aaron's breastplate after the children of Israel left Egypt in the 15th century BC (Exodus

28:17 and 39:10). The gem that is translated 'carbuncle' in the King James Bible (1611) is the Hebrew 'bareketh', or 'flashing' stone, after 'barak', meaning lightning bolt (Fernie, 1907, p. 158). However, throughout the ancient period 'baraketh' was uniformly associated with emerald (smaragdos) rather than carbuncle (Sinkankas, 1981, p. 84–85). It is recorded that the gemstones were to be donated by the Israelites, who had brought them out of Egypt.

Even if these gemstones were carbuncles, it is debatable whether or not they were garnets, since the red gems throughout antiquity were not separated on the basis of their physical or optical properties. However, if they were 'carbuncles', they were acquired in Egypt at a rather early time when gemstone use and demand was not widely developed. Therefore it seems most probable that they were garnets, possibly derived from early African sources, or from the early trade routes from India.

If the state of trading activity was localized and relatively undeveloped in the second millenium BC, there was much more development in the first millenium BC. Early in the first millenium Phoenician sailors explored the entire Mediterranean, west coastal lands of Africa, areas around the Red Sea, and even north to England. Trade colonies were established by them in many areas, including Carthage, peopled originally by merchant settlers from Phoenicia. This colony was to become important in early Roman times as a gem trading centre, mentioned by both Theophrastus and Pliny, and known particularly for red gemstones, probably garnet.

Also, in the same millenium, Persia moved armies into Europe in an ill-fated attempt to conquer Greece. The Greeks fought bravely at Thermopylae and finally stopped Xerxes' armies in a sea battle off Salamis. In the next century (4th century BC) the son of a Macedonian king returned the visit. Moving a disciplined army across many of the same roads crossed by Xerxes, Alexander moved into the heart of the Persian empire and defeated Darius in a pitched battle. Moving quickly, he crossed into India and subdued everyone in his path.

Although Alexander the Great died at the peak of his power, he opened up the riches of Asia to Greece and, eventually, Rome. Trade routes were already established prior to Alexander's times. But he succeeded in establishing an 'oriental' demand in Europe for the products of Asia resulting in an increased availability of red gemstones, and especially the garnets from India.

An early writer on gemstones, who was, incidentally, contemporary with Alexander, was Theophrastus. As a gem writer, he followed the pattern of Aristotle who attempted to explain all natural phenomena into four fundamental properties – air, water, earth and fire. The importance of Theophrastus is not only what he said but also when he said it. His treatise was written roughly about eight or nine years after Alexander's death in 323 BC, a date which proved to be a critical turning point for Greece and the West.

Theophrastus on garnet

The writings of Theophrastus on gemstones certainly was not the only discourse on the subject in the Greek period. But it was a major work, influencing later writers on the topic (including Pliny), and it was a comprehensive assessment of the subject. The years of Theophrastus (372–287 BC) span from the Classical period fo the Hellenistic, the latter ushered in by Alexander. After the time of Alexander, Greece was not to be the same. The new age altered its very existence. Art, religion and even its language began to undergo a change.

Trade began to expand and gemstones became much more available. Established trade routes by sea and overland encouraged consumerism. Gem centres developed in certain key areas of this trade cycle.

Theophrastus organized his discourse (*Peri Lithon*) around several topics. He talked about gem sources, colours, peculiar properties, uses, and types (species). He did not confine his observations to gemstones, however; he also discussed common rock types, including coal and pumice.

He classified the red gemstones into a type called 'anthrax', akin to the modern term anthracite (a type of coal). In his introduction to the anthrax gemstone, he happened to mention that it was 'incombustible', quite unlike charcoal of the same name. The reference to the incombustibility of anthrax seems to suggest that it was corundum, rather than garnet, since the former gem is incombustible with the blow pipe, while the latter is not. This passage probably influenced later writers, as Pliny also discussed the incombustibility of the anthrax gemstones (Pliny 37:25), causing later authors to speculate that the anthrax gems were corundum, not garnet.

However, in a modern translation of Theophrastus (Caley and Richards, 1956), the authors translated the Greek word (ακαυστον) rather loosely: 'But there is another kind of stone which seems to be of an exactly opposite nature, since it cannot be burnt'. That is, the anthrax gemstone, in contrast to a piece of charcoal (in the previous discussion) does not burn when the two stones are thrown into a common fire. Theophrastus is not saying that the anthrax gemstone is incombustible when tested by a blowpipe. He is simply marvelling over two 'rocks' called by the same name – one burns in a common fire, the other does not. The one that does not burn seems to burn from within (a paradox).

Unfortunately, Pliny and other authors misinterpreted this phrase to mean incombustible. Even Eichholz took the literal translation and had difficulty explaining the contradiction (Eichholz, 1965, p. 100).

That the anthrax mentioned in this passage was undoubtedly garnet and not corundum is confirmed by the reported use to which the gemstone was put. Theophrastus remarked that seals were cut from it. It is quite common knowledge that corundum seals were extremely rare even in Roman times, and unheard of in the Greek period. Indeed, in the extensive collection of seals from the British Museum, there are no examples of corundum seals from either Greek or Roman times, although the Museum has many garnet seals from both periods (personal interview 1984). Consequently the anthrax of Theophrastus must be taken as a garnet.

The anthrax was further described as red in colour, and appeared to be burning from within when held up to the sun, like a burning coal.

In a rare insight into the prevailing opinion of the day, Theophrastus remarked that the anthrax was quite valuable, as small carved specimens were selling for forty pieces of gold, a considerable sum in those days. Because of this high value put upon the gemstone, King could not accept this as describing as a 'common' garnet, but instead identified the gem as corundum (King, 1867, p. 225). However, it must be remembered that fine examples of any gemstone would have been a rarity in Theophrastus' day, and even garnet was not as common as it was in King's day, when the Bohemian mines were reaching their peak of production. Moreover, as Caley and Richards point out, the garnet seals were probably just being introduced into Greece at that time, and the novelty could have created an over-high price, particularly if the work was accomplished by a master engraver (Caley and Richards, 1956, p. 90).

The text of Theophrastus also reveals information on the sources of the anthrax gemstone. Carthage and Massalia (modern Marseilles) were mentioned as trading centres; anthrax gemstones were 'brought' from these centres. Both sites were early Phoenician trade colonies and products from anywhere in the known world might be found there. It could be speculated that garnets from Europe (Bohemia, in particular) might have been traded at Massalia, while African garnet sources were possibly found in the Carthaginian markets.

There are great difficulties in interpreting ancient mine sites from Greek and Roman authors who were not in the gem trade. Fictional mine sources may have been fabricated to conceal real sites. Actual sites worked, perhaps, for centuries might have been depleted, disappearing from modern geological research; known modern sources may have been unknown in ancient times. Moreover, the

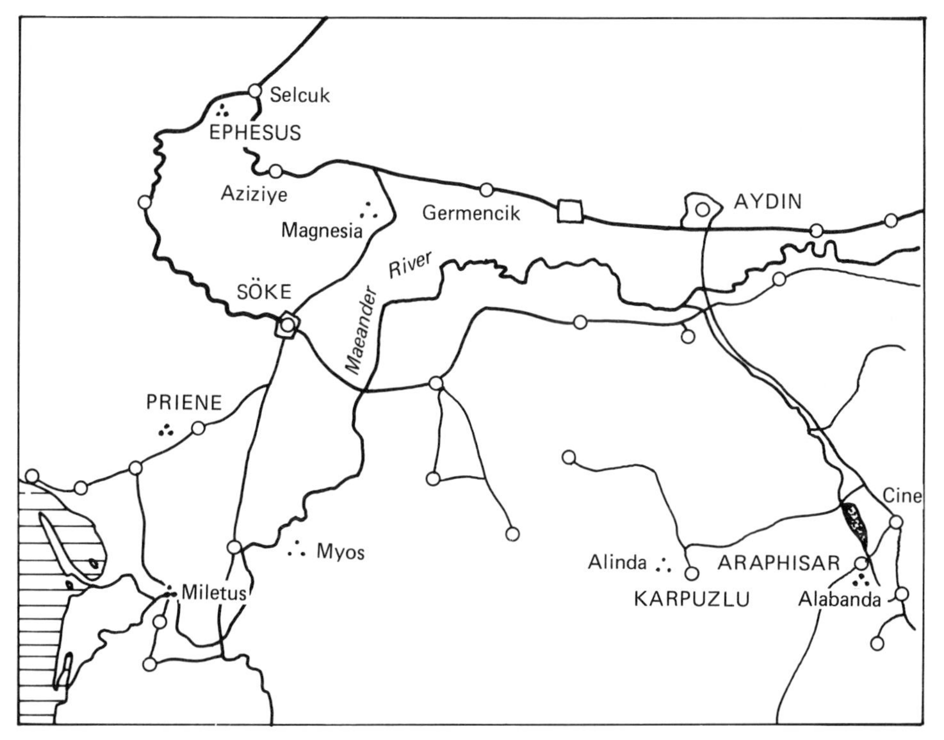

Figure 1.1 Map of area around Miletus. Ancient sites are marked with three dots. Note the relationship of Alabanda to Miletus

fragmentary knowledge of ancient geography could also hamper modern efforts in locating ancient mine sources. However, Theophrastus is quite explicit on Carthage and Massalia: they were trade centres, not mining sources.

In addition, in the next phrase, Theophrastus said that the anthrax also came from the area 'around' Miletus (see *Figure 1.1*). He did not explicitly say that they were mined there, but the thought could be implied. Since later writers (notably Pliny) also commented upon this site as a gem production and cutting area, it would be natural to assume that it was indeed a mining site.

Another possibility also exists for the Miletan site. Since it was strategically situated quite close to one of the major overland trading routes to India (the Ephesus to Antioch route, see *Figure 1.2*), it is conceivable that Indian gemstones, particularly garnets from northern India, were traded here. Perhaps both possibilities occurred. The few gemstones from the vicinity could have been consumed into the marketplace, depleting the local area, and resort was made to imported stones from India. Miletus could have also been engaged in sea trade from Antioch, as well as India, since it was a seaport town.

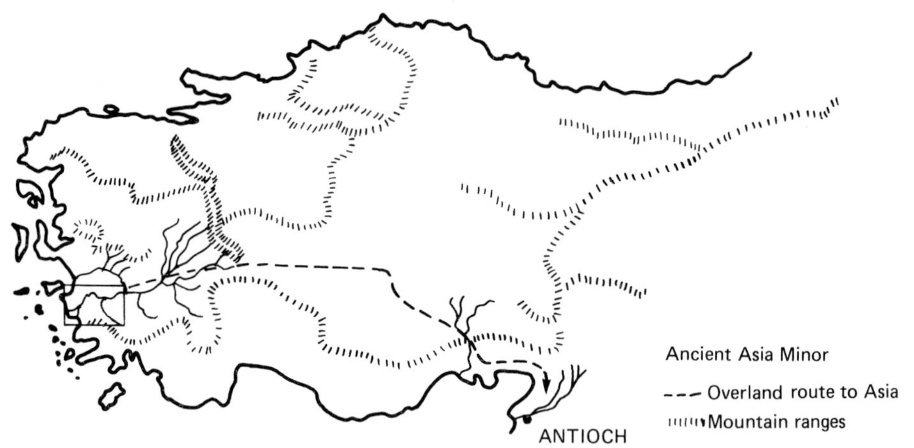

Figure 1.2 Map showing section of the overland trading route to India. The area in the rectangle on the left is shown in Figure 1.1

In his discussion of the Miletan anthrax, Theophrastus pointed out a peculiar phenomenon which did not seem to apply to the other anthrax stones – it was hexagonal in shape. Despite other authors' attempts to explain this stone as spinel (Caley and Richards, 1956) and almandine (Hill, 1774), it could only have been corundum because of this distinctive shape.

The presence of a hexagonal crystal in this locality seems to be a problem. All the other evidence seems to point to garnet as the anthrax of Theophrastus. Certainly this phenomenon cannot be explained away as a garnet feature, although by some stretch of the imagination if a dodecahedron garnet crystal is held in a certain way it could be made to reveal six angles. Nevertheless, the hexagonal crystal should be taken as a ruby rather than garnet.

However, quite probably several well-formed rubies were found in garnet parcels either from the area of Miletus, or carried there from India. Even the presence of a few well-formed crystals from a certain locality would be enough to arouse local interest and comment within the local marketplace. Such news quite possibly could have reached Theophrastus and other lapidary writers.

Also, because of the relative rarity of ruby in the modern world, it seems unlikely that the conditions in ancient Asia Minor (modern Turkey) were any different. Quantities of ruby crystals from a source as unlikely as Miletus seems most remote.

After Theophrastus described the anthrax gemstone, the narration digressed somewhat and then led to a discussion of other stones. However, he apparently returned to the subject of anthrax and enumerated 'anthrakion' from Orchomenos in Arcadia, Troezen and Corinth. Since the stones from Arcadia were large enough to be made into mirrors and they were colour banded, it is quite evident that he was no longer talking about garnet or even corundum crystals. So Theophrastus seemed to be describing three different stones under the classification 'anthrax'. The first type was charcoal, for which 'anthrax' was consistently used in the plural. Then there was the gemstone, which was for the most part, garnet and was singular and masculine in form. The third type, which may have been a form of onyx or obsidian was described in the neuter. By using anthrax in these three forms, Theophrastus could describe three separate stones.

Unfortunately Pliny missed this fine point and assigned Arcadia, Troezen and Corinth as carbuncle sources, an error that kept surfacing for the next 1900 years (for a recent reference, see Ball, *Historical Notes on Gem Mining*, 1931, p. 724).

In summary, all that can be deduced from the treatise of Theophrastus is that the 'fiery gems' were red. They were also intriguing when held up to the light, for they appeared to be burning, yet would not burn when thrown into a common fire with charcoal, a stone of the same species name. They were found in gem markets of Carthage and Massalia, as well as the Miletus area. A few stones of the Miletus source, whether a primary source or a trade centre, appeared to have been corundum, because of a hexagonal shape. Moreover, the gems were highly valued, as the cut signet stones could sell for as high as forty pieces of gold.

Although garnets were known to Theophrastus, he did not add to the nomenclature of the modern species. 'Pyrope' was not used at this time, neither was 'garnet' nor 'almandine'; even Alabanda was not mentioned directly by Theophrastus. Furthermore, Theophrastus set the stage for much later confusion; the use of a neuter form of anthrax being a prominent example. The lack of first-hand knowledge of gemstones and the gem market is painfully obvious. Yet Theophrastus should be given the credit for relating as much information as he did. He expounded on the nature, shape and colour of this gemstone, as well as the sources, prices, uses, and esteem to which the species were held. Significant also, was the total lack of superstition, lore and legend associated with the gem. There was no mention of amulets of anthrax nor medicinal uses for the gem. He did not suggest that anthrax had male and female counterparts (as Pliny and other writers did), although he was one of the first Western writers to divide gemstones into male and female counterparts.

From the time of Theophrastus to the time of Pliny, nearly four hundred years, there were many other writers on gemstones and lapidary. Most are lost; indeed, the only major writer on gems in the Roman period that survived into modern times was Pliny. Although he himself mentioned roughly 2000 other writers on the subject, only a few survived the destruction of the libraries and the chaos that followed the fall of the Roman empire. But this one major work survives and it provides one of the most important treatises on the subject of ancient mineralogy, profoundly influencing all later writers on the subject.

Pliny on the fiery gemstones

Caius Plinius Secundus lived in the golden age of the Roman Empire. He was born in 23 AD and died in the eruption of Vesuvius in 79 AD. His career was part military and part administrative; he was an avid reader and an energetic writer. His work on *Natural History* contained thirty-seven books on many subjects, from a mathematical survey of the universe to botany, anthropology, geography, zoology, and medicine. His last four books investigated minerals, fine arts and gemstones.

In order to fully appreciate the gemmological significance of Pliny's work, there must be some understanding of his times, and how they differed from those of Theophrastus. When Theophrastus wrote in 315 BC, trade with the East, although of ancient origin, was just developing due to the influx of Eastern products after Alexander's conquest. A period of consumerism, ushered in during the early Hellenistic period, grew in proportions as the Romans extended their empire in all directions in the first to the third centuries AD.

In Pliny's time, during the latter half of the first century AD, that trade extended east to India, south to Africa, and north to Europe. India was mentioned many times as a vital source for gemstones. Furthermore, the Roman taste for elegant materialism at home, despite legal attempts to control it, fuelled this trading activity.

In the preclassical Greek period, jewellery was mostly, if we can judge from many surviving examples, intricate goldwork without gemstones. In the Hellenistic period gemstones became more common, and jewellery often included them – particularly, cornelian, sardonyx and garnet – usually in that order of popularity. In Roman times, gemstones became more commonplace and a variety of different types were encountered, including ruby and diamond. Even a cursory reading of Pliny's carbuncle account, (carbuncle is the Latin equivalent to Greek anthrax) compared to the section in *Peri Lithon* by Theophrastus, indicates a dramatic change. A very discriminating consumer demand is evidenced, resulting in a much more complex gem marketplace and many more varied sources to satisfy that demand.

One of the major sources of carbuncle mentioned by Pliny was Carthage, or the Latin equivalent, Carchedonia, which was also described briefly by Theophrastus.

Carthaginian carbuncles

> *'Two kinds of 'carbunculi' are the Indian and the Garamantic: the latter was called in Greek the Carthaginian because it was associated with the wealth of Great Carthage'. (Pliny 37:25)*

The 'Garamantic' carbuncles were the Carthaginian anthrax gems of Theophrastus. This reference indicates a trading source that was in existence for nearly four hundred years, providing red gemstones to a gradually developing jewellery and fashion conscious public. These carbuncles were associated with Carthage, Pliny explained, because of the great opulence the city displayed in Greek times, indicating thereby that the gems were consumed in Carthage and not exported to other areas of the Mediterranean, as Theophrastus suggested.

Regarding the source, Pliny quoted a legend of the time which declared that these gemstones were mined in the mountains of the Nasamones – a people who lived near the Garamantes. Allegedly, the gems were formed 'by rains of divine origin', a phrase that Pliny reported from another author (37:30).

However, in an earlier book (Pliny 5:5), he revealed that another people, who lived to the southwest, were actually the original dealers who acquired the material deep in Africa. In describing the Amantes, who lived beyond the Nasamones, Pliny told of a group of cave-dwellers, 'the Trogodytas, with whom our only intercourse is the trade in the precious stone imported from Ethiopia which we call the carbuncle' (5:5). Very possibly the Ethiopian source was a trading centre as well, for Pliny described other Ethiopian stones at some length (37:25). It is quite conceivable that the source for the Carthaginian stones was India, transported perhaps from trade colonies in southern India to Ethiopia. However, Pliny seems unsure of his information at this point, for he relies heavily on secondary writers and seems to exhibit little first-hand knowledge of the African gem market. One thing seems certain, however: the Carthaginian mine source must have been rather extensive, lasting as it did for nearly four hundred years before Pliny, despite the destruction of Carthage (2nd century BC).

The Carthaginian carbuncles were further described as smaller than others from different sources, and dark in tone. The hue was purple 'in shadow', but flame-red in transmitted daylight, suggestive of either garnet or perhaps dark corundum. Furthermore, the Carthaginian stones sparkled when they are held against the sun; the latter description is indicative of their method of viewing the stones and evaluation techniques. The 'sparkling' effect was probably due to inclusions. That the Carthaginian stones were essentially garnets, not corundum, is suggested when Pliny quoted Archelaus that the Carthaginian gemstones were used as signets (37:25).

Pliny divided the Carthaginian carbuncles into two classes, male and female, following the ancient traditions as used by Theophrastus (but not for his anthrax). In Pliny's day this division was a quality grading system, for his 'male' stones were darker and more 'brilliant', while his female stones exhibited a lighter tone, and were considered weak (languid).

Any of the various carbuncles mentioned by Pliny could have had male or female characteristics, indicating fine or poor qualities, regardless of source. In comparing the various types and sources of the carbuncle, however, Pliny reported that the Carthaginian gems were 'far less valuable' than the rest (37:30).

It is possible that one form of the male Carthaginian gemstone was asteriated, perhaps a star garnet. This idea is suggested by the statement that the male Carthaginian stones had a 'blazing star inside them . . .' (37:25). However, it was most likely an inclusion within the stone that 'sparkled' like a star since the Latin word 'intus' was used (meaning 'from within'), indicating that the star was seen from within, and not on the outside surface.

Carbuncle from Asia Minor

Another major source of carbuncles mentioned in Pliny's account was Caria in Asia Minor (modern Turkey), including the towns of Miletus, Orthosia and Alabanda. Since this area was also mentioned by Theophrastus (Miletus), it can be assumed to have been an ancient source or gem market.

Pliny also mentioned that in this district a very 'poor' variety of rock crystal was produced (37:9), and was an unknown type of black fusible rock (he may have been referring to low quality carbuncle here, however: see 36:13), in addition to the carbuncle mined at Orthosia (37:25), but cut at Alabanda. It seems likely that the area was gem producing, though there are neither modern gemstone sources, nor evidence of ancient gem mining activities. However, an extensive emery deposit is reported (C. Schmeiszer, *Zeitschrift fur praktische Geologie*, XIV, 1906, p. 188, as cited by Caley and Richards, 1956). Since emery contains corundum, in this case about 50%, it would have been invaluable as a lapidary cutting agent, and was undoubtedly used in the cutting shops in Alabanda.

The reference to Alabanda as a cutting centre, a fact not mentioned by Theophrastus, seems to indicate at least a minor industry developed there in the four hundred years between Theophrastus and Pliny. No doubt the cutters included gem engravers who cut intaglios for signet rings. Such gemstones included cornelian, sardonyx, garnet as well as pastes and other materials. Images were carved in negative relief; almost always the gems were ringstone size (ca. 18 × 13 mm) and the carvings were intricate sculptures in miniature. The owners of such ringstones would 'sign' official documents and letters in sealing wax. It is well known that most Romans treasured their seal-rings, which often reflected aspects of the owner's personality. Kunz (*Rings for the Finger,* p. 127) related the story of a Roman citizen of Sicily in the first century BC who was forced to give his signet ring up to the Roman governor of the area because the governor was impressed with the design. Kunz commented that 'the injustice' of this act must have been felt all the more keenly because the special and peculiar design on a seal was then regarded as something closely linked with the personality of the owner.

The importance and the universality of the seal rings in Roman times is unquestioned. The fact that garnet was a popular medium for intaglios, suggested that Alabanda formed an integral part of a major Roman industry, and was probably a garnet gem centre as well.

In describing the appearance of the carbuncles from Alabanda and Orthosia, Pliny reported that they were 'darker than the rest, and rough' (37:25). Another type of carbuncle was also reportedly found in the area – the prized 'Lychnis' carbuncle. It was called by that name because its beauty was especially apparent under lamp light (37:29). It's colour varied from purple to red, the second grade of which resembled the so-called 'Flower of Jove', ('Agrostemma flos Jovis', as described in Pliny, XXI, 59, 67).

Lychnis was found in sizes large enough to fashion drinking vessels. This reference provides evidence to suggest that the Lychnis gemstones were large garnets, although Eichholz took the Indian Lychnis stones as possible corundum (Eichholz, Pliny, p. 247, note e).

While the Lychnis gemstones reportedly occurred in Caria, they also were to be found from India. Indeed, the finest specimens were said to be Indian, not Carian (37:29). Very likely they represented variations in colour: perhaps tonal and intensity fluctuations which are quite common in either corundum or red garnets. Like the Carthaginian stones, the Lychnis gems exhibited a pyro-electric magnetism, a condition possible with garnet or tourmaline, but not corundum.

Carbuncles from other locations

In addition to carbuncles from either Carthage or Alabanda, which were apparently the major centres in Roman times, Pliny also described other centres. Ethiopia was mentioned as a source, as was Egypt (near Thebes). Very little is mentioned about stones from these areas, except that the Egyptian stones were brittle, full of veins, and often carved into drinking cups; the Ethiopian stones 'looked greasy and shed no lustre at all'. Probably both sources were trading stations and not mining areas. Lisbon was also reported as a gem location, Pliny quoting Bocchus for this source. Massilia was also mentioned, but Pliny had to quote Theophrastus for this information (Pliny 37:25). It seems possible that the Massilia source either ceased to function as a gem export centre perhaps after the demise of Carthage, or it was unknown to Pliny. It also seems strange that Pliny was not aware of the Bohemian mines, which could have been active during this period. Sometimes Pliny pitted one source against another. He quoted many writers' opinion that Indian carbuncles were brighter than the Carthaginian. Yet one writer (Satyrus) asserted that Indian carbuncles lacked brilliance and were generally flawed. Pliny remained aloof; he just gathered the data and presented it, often without evaluating its substance or consequence.

In one such instance Pliny misread Theophrastus and added Orchomenos in Arcadia, Chios, Troezen and Corinth as additional carbuncle sources (see discussion under Theophrastus). Theophrastus called these stones by the term 'anthrakion', the neuter form, with a description that could not have been 'anthrax'. Eichholz tentatively identified the stones from Orchomenes and Chios to be 'dark marble' and the stones of Troezen and Corinth as serpentine or a variegated marble (Theophrastos Eichholz, VI:33, footnote, p. 111). Pliny apparently mistook 'anthrakion' for 'anthrax' and included the section from Theophrastus almost verbatim.

Carbuncle gender classification and quality analysis

Throughout his book on gems, Pliny does not hesitate to give information on fashion preferences. He also did so for the carbuncles. The highest order of gems was attributed to the male series. Carbuncle was divided into male and female species, a common practice in ancient times. Indeed, the practice dated back to the ancient Babylonians (3000–2000 BC), and was carried on by the Assyrians, the Chinese, the Hindus, and modern day Arabians, Negro tribes of the upper Nile, peoples of the Caucasus, and the Zuni and Navajo Indians (Ball, *Roman Book*, p. 276, footnote #6).

The finest carbuncles were male, characterized by exhibiting more brilliance and producing a darker, richer colour. The optimum colour was found in a type called the 'amethystizontas', which were described as 'those in which the fiery red shade passes at the edge into amethyst-violet'. A second variety, the 'Syrtitae', are said to be second best. They radiated a 'feathery' brilliance and one so bright that they 'reveal themselves in ground where sunlight is reflected most powerfully'. (Pliny, 37:25).

Callistratus is quoted as saying that carbuncle displays a dazzling brilliance, 'so that when placed on a surface it enhances the lustre of other stones that are clouded at the edges'. The idea seems to be that the finest gem, when placed beside others of much less clarity, would enhance the beauty of the others as well, due to its dazzling effects. Consequently, Callistratus recalled that people referred to such beautiful carbuncle gems as 'bright' carbuncles, as opposed to the 'cloudy' carbuncles (Pliny, 37:25).

It is quite evident from the previous passage that the Romans intently studied the effects of the colour under transmitted light and compared samples. Those which seemed to generate the glistening or dazzling effects of a rich colour were highly valued. Included stones, unless they were of a special sort, were less desirable and exhibited a 'cloudy' brilliance. The 'Lychnis' stones apparently reacted favourably under lamp light: the colour was not too dark nor too light, with a pleasing intensity and hue. However, despite their exceptional beauty, they were somewhat inferior to the finest carbuncle because they were more difficult to engrave.

Therefore, Roman tastes were rather well developed around colour, brilliance, and clarity. Although it is sometimes confusing to extract Roman viewpoints in different situations, their quality analysis seems rather modern.

False carbuncles in Roman times

In addition to providing much data on the description, sources and quality analysis of the carbuncle gemstones, Pliny also reported gemmological tests designed to separate genuine carbuncles from the fakes.

Modern curiosity is aroused when Pliny reported the examples of enhancement techniques found in the trade. One trick to make a nearly opaque carbuncle appear translucent was by placing a foil or some other substance underneath the gemstone. Although the origin of foil-backed stones is very old (see Ball 1950, ch. X) it is interesting to find it in use here. Another trick was to soak 'dull' stones in vinegar for fourteen days. The result would produce a 'brilliance' that would last for months.

Counterfeit carbuncles were formed in glass. The glass stones, however, were detected by touching to a grindstone (glass, being softer, was also brittle and was easily detected in the process). Another, less destructive test, was by weighing, presumably with a known stone of nearly the same size as the sample, for glass was known to weigh less than carbuncle of the same relative size. The 'specific gravity' test therefore, if it did not originate in the Roman period, was certainly utilized here.

Another type of counterfeit stone in use was a glass with inclusions similar to carbuncle. These stones were apparently somewhat more difficult to detect once the gem was mounted, for suspicions of glass would have been allayed by the presence of the inclusions. Obviously, glass was created in the same colour range as the gemstone, and the 'inclusions' were probably created by 'heat-cracking' the glass, inducing fractures similar to cracks in natural gemstones.

Evaluation of Pliny on carbuncle

Despite his uncritical character, Pliny did provide a rare glimpse of gemmology in the Roman period. He quoted authors that he was familiar with on the subject, provided intriguing information on gem quality analysis, and gave as much data on the gem sources as he was capable of finding out. There was a rationalism in his approach; he was not given to provide magical formulas regarding the gemstones. He often prefaced some story about the efficacy of a stone with, 'it is said'. He was sceptical of eastern magic but he did, however, in the same vein as Theophrastus, give medicinal remedies associated with the gemstones.

In his discussion of 'carbuncle', he introduced the subject by promising to expound upon all the 'flaming' gemstones found in nature. He divided the carbuncle into one class differentiated by source; another class was divided into male and female, irrespective of source.

Some of the sources are familiar to readers of Theophrastus – Carthage, Miletus, and Massilia. Others are not – Ethiopia, Alabanda (no doubt the same source of nearby Miletus mentioned by Theophrastus), Egypt, and Arabia.

Pliny's weakness is that his knowledge of the gem industry is either secondhand, or he simply relied on prevailing general knowledge. He seemed to have little first-hand knowledge gained by personal investigation. Consequently, he was sometimes unaware of sources, whether they were trade centres or mining sites. But he did mention the existence of fake gemstones and tests available for detection; and he also discussed treatment processes.

In sheer volume, also, Pliny overshadowed Theophrastus. Whereas the entire anthrax section of Theophrastus can be compressed into one paragraph, Pliny devotes six chapters to the subject.

It is quite apparent that most of the carbuncle described by Pliny was garnet. There may have been some ruby and spinel included in the category at this period, mostly brought in from the Ceylon mines. But the major spinel finds of Badakshan (Afghanistan) may not have been in production at this early period. Ball (1931, p. 720, 721) reported the earliest reference to the mines in 951 AD although it was possibly known earlier. Moreover, it is questionable whether the red tourmalines of Africa were known in Pliny's day, but the lack of surviving samples suggest they were not.

Later lapidary writers

Several other writers wrote about gemstomes in the later Roman period. One such author was C. Julius Solinus who wrote *Collectanea Rerum Memorabilium*, dated variously between the 2nd and the 4th century AD. Solinus wrote in the same vein as Pliny and was little influenced by the magic and the superstitition from the Alexandrian school which was by now gaining ground since its beginning in 148 AD. However, the account was not a treatise on minerals or gemstones, but a rambling geographical summary listing products from each area. Unfortunately, Solinus relied heavily on Pliny for much of his information, so little was new.

Solinus, however, did use the term 'pyropus' in describing the gemstone 'ceraunio'. Unlike Pliny, Solinus described the colour of the ceraunio gem to be 'fire-like' (pyropus). It seems that the term was a general word that retained the elements of its original meaning from the Greek period. Nevertheless, it may have been from this and perhaps other Latin or Greek references, that later writers used the term to describe a variety of garnet. Oddly enough, the ceraunio stone was described by Pliny to be a bluish colour, perhaps our moonstone, rather than a type of carbuncle.

Carbuncle was only mentioned incidentally in connection with its source in Africa (Trogodytis and the Nassamones). Nothing was mentioned concerning its qualities, types or other sources, with the sole exception of the 'lychniten' which was derived largely from Pliny. Hyacinth was also described in the same vein as Pliny: a blue gem, probably sapphire, which varied slightly in colour towards the red.

The writings of Damigeron are very similar to those of Solinus. Damigeron was perhaps a Hellenistic writer (Evans, 1970, p. 20), whose works survive only in Latin translations, dating from the second century AD (or later). In his work he listed very brief accounts of fifty stones. Carbuncle was not in the list, but there was one stone familiar to the readers of Pliny, that of the Lychnites, perhaps identical to Pliny's lychnis, said to have been consecrated to the deity Vulcan.

Epiphanius, writing from the Christian standpoint, offers little more than a garbled version of Pliny. He did mention that the anthrax or carbuncle is difficult to steal, 'because if he (the thief) conceals it inside his clothes, the glow (of the stone) shines out through the folds' (Maxwell-Stuart, 1977, p. 436).

In the interim between the fall of the Roman Empire and the 11th century writings of Marbode, western traditions of the ancient world were kept alive through many Arabic translations and writings. Two writers were referred to often by the Medieval writers. One was named Evax, who might have been an author during the time of Tiberius in the first century AD (Evans, 1970, p. 20, note 6). Little else is known of him, although several of the Medieval writers dedicated their work to him, and claimed to write in the same tradition initiated by him.

Another author quoted frequently was Aristotle. However, the famous Greek philosopher was not known to write on stones. Possibly this author was an Arabic writer either named Aristotle, or one using the pseudonym. Evans (1970, pp. 38–39) quoted the compiler of the Lapidary of Alfonso X, who described aspects of 'Aristotle's' work. That work described seven hundred stones in terms of their qualities, colours, sources and their virtues. Unfortunately the work is lost but it was known to several Medieval writers.

Marbode wrote a treatise on lapidary in verse, published sometime before his death in 1081 AD. An English translation is found in one of King's works (1866).

The work was based largely on that of Damigeron in style, but influences are also noticed from other sources, perhaps those of Evax and the pseudo-Aristotle. For the first time garnet is mentioned as a variety of hyacinth, and 'alabandine' seems to have been established as a gem species that competed well with sard. Even carbuncle was described as having twelve varieties with mines in the Lybian desert (the Carthaginian stones of Pliny). Magic and medicinal remedies are frequently mentioned for each stone. The term garnet was taken from the pomegranate seeds, whose colour was similar to the colour of one of the varieties of hyacinth (red).

Another writer of the Middle Ages was Albert the Great (1206–1280 AD). Astrological influences are seen in Albert's writings of the various gemstones, but several new topics are observed. First, 'belagius' was used to describe a type of carbuncle, perhaps derived from pseudo-Aristotle. Belagius was attributed by Albert to the term 'palatium' which was said by Albert's contemporaries to represent the 'house' of the carbuncle. In addition to belagius, Albert also classified rubinus and granatus as species of gemstones, setting the stage for three categories of red stones (spinel, ruby and garnet) which would finally emerge much later. Despite Albert's explanation for the derivation of 'belagius', it has been suggested that the term was derived from 'Badakshan', the source of both ruby and spinel which was active by the tenth century (Wyckoff, 1967, p. 75; the modern term of 'belagius' is 'balas', equated with spinel).

In 1520, Camillus Leonardus, a medical doctor, also wrote on stones and their virtues. Although he quoted Albert and Marbode in suggesting twelve varieties of carbuncle, he only listed five. The first was the most noble, carbuncle; the others, in order of their power and value, were ruby, balasius, spinella and granate.

Later in the same century, George Agricola wrote his landmark work on gemstones. In it, he described new sources for the carbuncle, including the Bohemian garnet deposits in Czechoslovakia (Agricola, 1546, p. 172). He also explained that the spinel derived its name from small red stones, since it was used in the diminutive sense ('spinellus'). In 1609 Boetius de Boodt's important work also included a new source of garnet from Pegu, a kingdom in southern Burma, as well as the ancient sources. Garnets from Bohemia were analyzed for their quality and price and many categories are listed.

Nicols (1652) drew heavily from de Boodt. However, he divided the types of carbuncle into the categories of ruby, garnet, almandine and red hyacinth. 'Balassius' was elevated to a level with carbuncle, but there is some confusion whether spinel was considered part of balassius or if it represented a separate category.

Of interest in all these works are the frauds that were described in their connection. The main difficulty with classification in this period was a lack of understanding of the chemical composition of the various gemstones. Without such knowledge, gem enthusiasts of the period could only rely on colour, hardness, specific gravity and crystal form for identification and classification tasks.

The science of mineralogy emerged in the late eighteenth century, with the works of Haüy (1780), Lavoisier (1789) and others, finally bringing to a close an era, thousands of years old, based on gemstone identification by observation and limited physical testing. It was an era steeped in metaphysics and tales of medicinal remedies and special powers. The romance of gemmology somehow lost its special charm; one work based on this new science (Mohs, 1825) is tedious and difficult to read. The magic is gone, and the charm is lost. But the new age did not bring the old one to an abrupt end: as late as 1907, medicinal remedicies were still being promoted (Fernie, 1907).

The contributions of the new age of gemmology can be studied in the monumental works by Jamieson, Mohs, Feuchtwanger, and several generations of Danas. The emergence of garnet into the modern species is largely due to the application of the tools of the new science by these leaders. Certainly Bauer's landmark work on gemstones was due to the contributions of these early writers and researchers. At the beginning of the twentieth century, the essential six species of garnet were already in place. Research in the twentieth century discovered complexities well beyond the understanding of the late nineteenth century writers. Garnet is an incredibly complex mineral that is full of surprises for the scientist, the student of gemmology and the layman. But the new age is very young, and the thrill of discovery is the hallmark of the modern age.

Bibliography

AGRICOLA, D., *de Natura fossilium,* Libri X, 1546, translated into German by Georg Fraustadt, Berlin (1958)

AGRICOLA, GEORGIUS, *De Natura fossilium,* English translation by Mark Chance Bandy and Jean A. Bandy, Nov., 1955, by the Geological Society of America, Special Paper 63

"Alabanda", *Oxford Classical Dictionary,* 2nd Edition, Oxford (1970)

ALBERTUS MAGNUS, Book of Minerals, translated by Dorothy Wyckoff, Oxford (1967)

BALL, SYDNEY H., 'Historical notes on gem mining', *Economic Geology,* Vol. XXVI, No. 7 (Nov., 1931)

BALL, SYDNEY H., *A Roman Book on Precious Stones and English Modernization of the 37th book of the History of the World by E. Plinius Secundus,* Los Angeles (1950)

BAUER, MAX, *Precious Stones,* translated by L. J. Spencer, Rutland, Vermont and Tokyo, Japan (1970). Reprint of the 1905 edition

BEAN, G. E., 'Alabanda', *The Princeton Encyclopedia of Classical Sites,* Richard Stillwell (ed.) Princeton, New Jersey (1976)

BERQUEN, ROBERT DE, 'De L'iris, La Vermeille, Escarboucle ou Grenat, and de la Cornaline', *Marchaud Orpheure a Paris,* Paris (1560)

BOARDMAN, JOHN, *Greek Gems and Finger Rings,* London (1970)

BOODT, BOETIUS DE, *Gemmarum et Lapidum,* 1609, books I and II

BURY, SHIRLEY, *Jewellery Gallery Summary Catalogue,* London (1983)

'Damigeron', *Paulys Real-Encyclopadie der Classischen Altertumswissenschaft,* Stuttgart, p. 2055 (1901)

DAMIGERON, *Orphei Lithica, accedit Damigeronde Lapidibus recensuit,* Eugenius Abel, Berolini (1881)

DANA, EDWARD SALISBURY, *The System of Mineralogy of James Dwight Dana (1837–1868),* 6th edition, London (1896)

EVANS, JOAN, *Magical Jewels,* Dover Publications Inc., New York (1976). Reprint of the 1922 edition

FARRINGTON, OLIVER CUMMINGS, *Gems and Gem Minerals,* Chicago (1903)

FERNIE, W. T., *Precious Stones: For Curative Wear; and Other Remedial uses. Likewise the Nobler Metals,* Bristol (1907)

FEUCHTWANGER, L., *A Popular Treatise on Gems,* New York (1867)

FURTWANGLER, ADOLPH, *Die antiken Gemmen, Geschichte der Steinschneidekunst im klasischen Altertum,* Leipzig, Berlin, 1900. 3 vols

GARHARD, CAROLUS ABRAHAMUS, 'Disquito physico-chymica Granatorum Silesiae atque Bohemiae', *Opuscula Mineralogica,* **10** (1760)

HIGGINS, R. A., *Greek and Roman Jewellery,* London (1961)

KING, C. W., *The Natural History of Precious Stones, etc.,* London (1865)

KING, C. W., *Antique Gems: their Origin, Uses, and Value,* London (1866)

KING, C. W., *The Natural History of Precious Stones,* London (1867)

KING, C. W., *Antique Gems and Rings,* London (1872)

KLEINER, G., 'Miletos', *The Princeton Encyclopedia of Classical Sites,* ed. by Richard Stillwell, Princeton, New Jersey (1976)

KUNZ, GEORGE FREDERICK, *Rings for the Finger* (1973) Dover. Reprint of the 1917 edition

LEONARDUS, CAMILLUS, *The Mirror of Stones,* transl. into English (1750) from the original 1520 Latin edition

MARSHALL, F. H., *Catalogue of the Jewellery, Greek, Etruscan and Roman in the Department of Antiquities, British Museum,* London (1911)

MARSHALL, F. H., *Catalogue of the Finger Rings, Greek, Etruscan and Roman in the Department of Antiquities, British Museum,* London (1907)

MAXWELL-STUART, P. G., 'Epiphanius on Gemstones', *The Journal of Gemmology,* October (1977)

MIDDLETON, J. HENRY, *The Engraved Gems of Classical Times, with a catalogue of the gems in the Fitzwilliam Museum,* Cambridge (1891)

MOHS, FREDERICK, *Treatise on Mineralogy,* Edinburgh and London, 3 vols. (1825)

MOORE, N. F., *Ancient Mineralogy,* New York (1834)

NICOLS, THOMAS, *A Lapidary or, the history of pretious stones: with cautions for the undeceiving of all those that deal with pretious stones,* Cambridge (1652)

OGDEN, J. M., *Jewellery of the Ancient World: Materials and Techniques,* London (1982)

PLINY, GAIUS PLINIUS SECUNDUS, Naturalis Historia, transl. by D. E. Eichholz, Loeb Classical Library (1971)

RICHTER, GISELA, M. A., *Engraved Gems of the Greeks and the Etruscans,* 2 vols., Phaidon, London (1968)

SINKANKAS, JOHN, *Beryl,* Van Nostrand (1983)

SOLINUS, C. JULIUS, *Collectanea Rerum Memorabilium,* Th. Mommsen, Berolini (1864)

TAYLOR, GERALD and SCARISBRICK, DIANA, *Finger Rings from Ancient Egypt to the Present Day,* London (1978)

THEOPHRASTUS, *On Stones,* transl. by Earle R. Caley and John F. C. Richards, Columbus, Ohio (1956)

THEOPHRASTUS, De Lapidibus, transl. by D. E. Eichholz, Oxford (1965)

THEOPHRASTUS, History of Stones, transl. by John Hill (1774)

WALTERS, H. B., *Catalogue of Engraved Gems, Cameos, Greek, Etruscan and Roman, in the British Museum,* London (1926)

WARMINGTON, E. H., *The Commerce Between the Roman Empire and India,* Cambridge (1928)

Chapter 2

Modern garnet

Modern gemstone classification is based upon species and varieties (see *Table 2.1*). For example, corundum is a species category; 'ruby' and 'sapphire' are variety names of the corundum species. In garnets, however, the classification is somewhat more complex. The term 'garnet' is a group name, while the so-called end-members are species (almandine, pyrope, spessartite, andradite, uvarovite, grossular, etc.).

TABLE 2.1. Group, species and varieties of the garnet family of minerals

Group name	*Species name*	*Trade names*	*Variety names*
GARNET	Pyrope		
	(Pyrope-almandine)	'Rhodolite', 'Malaya'	
	Almandine		
	(Spessartite-almandine)		
	Spessartite		
	Grossular		Hessonite; Tsavorite
	Andradite		Demantoid; Topazolite
	Uvarovite		
	Khoharite		
	Knorringite		
	Yamatoite		
	Calderite		
	Blythite		
	Goldmanite		
	Kimzeyite		
	Ferric-kimzeyite		
	Skiagite		
	Yttrogarnet		
	Hydrogrossular		
	Hydroandradite		
	Schorlomite		

Gem species have traditionally been limited to pyrope, almandine, spessartite, grossular, hydrogrossular and andradite, but many very rare end-members have been found as mineral specimens, and their presence can sometimes be used to understand complex compositions of unusual gem garnets.

Gem garnet varieties are hessonite and tsavolite for the grossular species; and demantoid and topazolite for the andradite species. Occasionally rare end-member garnets will be thought of as varieties of the major species; however, in our definition, the end-member is equivalent to species.

In addition to group, species and variety names, garnets are sometimes referred to as a series of species. For example, 'pyralspite' is often used to represent the species pyrope, almandine and spessartite. 'Ugrandite' is another series term, referring to the species uvarovite, grossular and andradite in a collective sense. Even two species may be referred to as a series category. 'Pyrandine' (Pyrope-almandine), 'spandite' (spessartite-andradite) and 'grandite' (grossular-andradite) are examples of such adaptations. However, the group, species and variety names represent the major classification terms of the garnet family.

Trade names are usually controversial, especially within the science of mineralogy. Nevertheless, the trade terms 'Rhodolite' and 'Malaya' are included in *Table 2.1*, since they are widely used (and misused) in the gem trade. Wisely, such historical terms as 'Arizona ruby' or 'Elie ruby' and many others have been dropped from modern trade usage as the study of gemmology (and mineralogy) has penetrated jewellery stores around the world, dictating standards, testing procedures, classification terminology, as well as simple ethics to the industry.

Names for gemstones continue to be fabricated by promoters who are anxious to arouse consumer emotions to create or stimulate sales. But the industry as a whole is much more sensitive to deceptive practices, and consumer awareness of gemstones through books and educational institutions reinforce these standards.

Garnet chemistry

One of the most uniform garnet characteristics is its chemical formula. The basic chemical formula for garnet can be expressed as $A_3B_2C_3$ (see *Table 2.2*). The 'A' can be magnesium (pyrope, khoharite, and knorringite), manganese (spessartite, yamatoite, calderite and blythite), iron (almandine and skiagite) or calcium (grossular, andradite, uvarovite, goldmanite, the kimzeyites and the hydrogrossulars). The 'B' can be aluminum, iron, vanadium, chromium or titanium, while 'C' can be silicon, iron, aluminum or titanium.

The above arrangement is seldom, if ever, composed of just three elements, for each of the three positions may include several elements, although one is usually dominant. For example, an almandine may include a certain proportion of magnesium, replacing the iron of its bulk chemistry. This replacement is called 'isomorphous replacement', and it provides the key to understanding garnet composition. Alterations in the 'B' and 'C' positions create unusual garnet compositions, but the basic structure for the gem material is an aluminum or iron silicate with the 'A' composed of either magnesium, manganese, iron or calcium, or combinations thereof. Chemical analyses have indicated a slight variation in the ideal structure of 3:2:3, although it is generally consistent.

With such a number of composition possibilities, garnet species could conceivably produce as many as sixty end-members (Manson and Stockton, 1981).

TABLE 2.2. Chemistries of the garnet end-members, including the rare species

GARNET CHEMICAL COMPOSITION

BASIC FORMULA: $A_3 B_2 C_3$

A = Ca, Mn, Mg, Fe
B = Al, Fe, Ti, V, Cr
C = Si, Fe, Ti, Al

PYRALSPITE SERIES	UGRANDITE SERIES
Pyrope: $Mg_3 Al_2 (SiO_4)_3$	Grossular: $Ca_3 Al_2 (SiO_4)_3$
Almandine: $Fe_3 Al_2 (SiO_4)_3$	Andradite: $Ca_3 Fe_2 SiO_4)_3$
Spessartite: $Mn_3 Al_2 (SiO_4)_3$	Uvarovite: $Ca_3 Cr_2 (SiO_4)_3$

OTHERS

Goldmanite: $Ca_3 V_2 (SiO_4)_3$	Calderite: $Mn_3 Fe_2 (SiO_4)_3$
Yamatoite: $Mn_3 V_2 (SiO_4)_3$	Blythite: $Mn_3 Mn_2 (SiO_4)_3$
Khoharite: $Mg Fe_2 (SiO_4)_3$	Kimzeyite: $Ca_3 Zr_2 (Al_2Si)O_{12}$
Knorringite: $Mg_3 Cr_2 (SiO_4)_3$	Ferric-Kimzeyite: $Ca_3 Zr_2 (Fe_2Si)O_{12}$
Skiagite: $Fe_3 Fe_2 (SiO_4)_3$	Hydrogrossular: $Ca_3 Al_2 H_{12} O_{12}$
Yttrogarnet: $Y_3 Al_2 Al_3 O_{12}$	Hydroandradite: $Ca_3 Fe_2 H_{12} O_{12}$

However, although over twenty are known to exist, they are very rarely encountered as mineral specimens. Six are usually important for garnet classification and only five exist in commercial quantities (pyrope, almandine, spessartite, grossular and andradite). The balance have not been found to exist, or they have not occurred in gem qualities. Nevertheless they are treated in this text because knowledge of their existence is sometimes necessary to explain abnormalities in gem garnets. Also, possibilities for future gem deposits within these rare species cannot be ruled out.

Gemmologists do not use chemical analysis for routine laboratory testing. However, the chemical variations observed above produce changes in the physical and optical properties of the garnet specimen. Gemmological tests reveal these changes and, usually with some degree of certainty, the correct species and variety can be properly assigned to the specimen.

In addition to the complexity of the chemical composition, garnets also exhibit miscibility characteristics. Certain garnet species are said to be in 'solid solution' with the other if a high miscibility relationship exists between them. Garnets in solid solution with each other may produce any proportion of garnet between the two end-member species. For example, pyrope-almandine mixtures can exist wherein the ratio of Mg/Fe can be 90/10, 80/20, 70/30, 60/40, 50/50, 40/60, 30/70, 20/80, 10/90, or any combination in between these proportions. When there is a large category of gem samples in between the two end-members, then an intermediate classification is adopted. The miscibility relationships of the various major garnets are illustrated in *Figure 2.1*. The garnets in solid solution with each other are pyrope-almandine, pyrope-spessartite, almandine-spessartite, grossular-andradite and uvarovite-andradite.

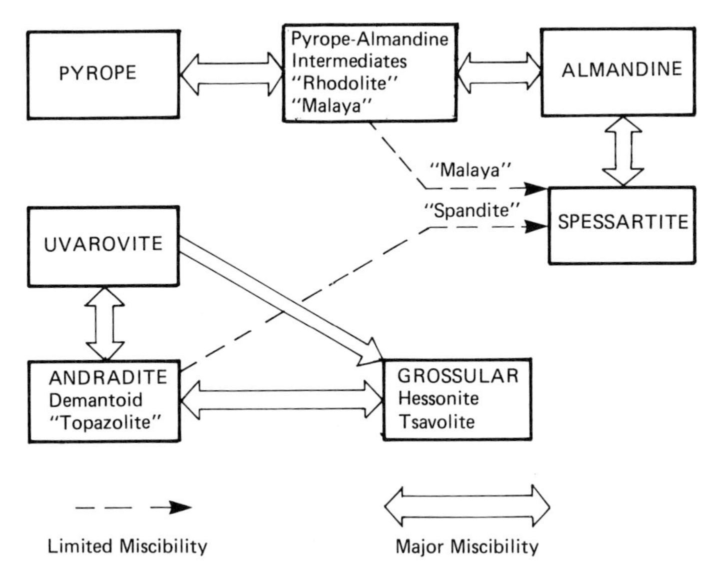

Figure 2.1 Miscibility relationships with the major end-members of the garnet family

There are also possibilities of solid solution between grossular and spessartite (see Chapter 6) and spessartite-andradite ('spandite'). Limited miscibility exists for most of the gem garnets, although gem andradites in both demantoid and topazolite varieties usually produce very high andradite end-member purity ratios (to 99.67%) thereby limiting the miscibility (Manson and Stockton, 1983). Grossular also can exhibit high end-member purities (to 96.6%), but they are less common than the andradites, and less consistent.

It is quite evident that the identity of the individual garnet is dependent upon its chemical structure. In the times before chemical analysis was known, it is no wonder that lapidary writers could not properly define garnets, especially. Observation alone, supplemented by specific gravity and hardness tests could not have clarified this complex classification problem.

Garnet crystallography

Very often garnet is found in alluvial deposits in which the crystal faces are obliterated. This is another reason why the ancient gem writers and enthusiasts could not properly identify the material. Yet there are many examples of garnet crystals occurring in nature. They form in the cubic system, although cubic faces are rarely encountered. The most common garnet crystal form is the dodecahedron (literally, 'twelve faces'), which is so often found that it is called the 'garnetohedron' (*Figure 2.2a*). It exhibits twelve faces with four sides. Another common form is the trapezohedron (literally, 'twenty-four faces'), consisting of twenty-four faces and eight sides (*Figure 2.2b*).

Sometimes the two basic forms are combined, where either the dodecahedron faces ('d') or the trapezohedron faces ('n') dominate (*Figure 2.3*). Less common

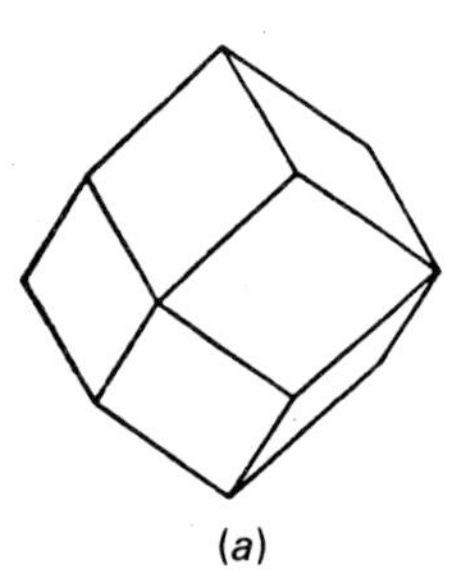

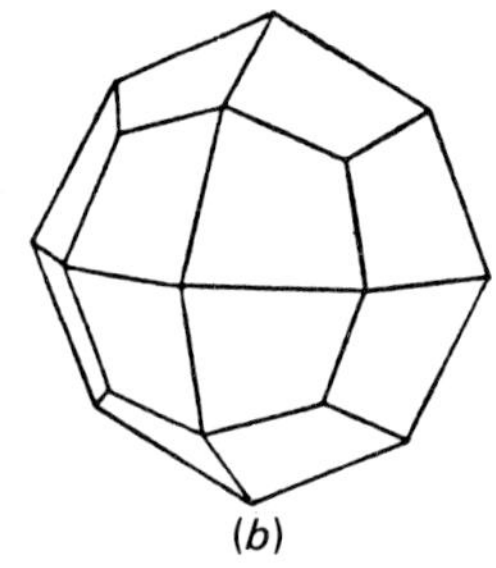

Figure 2.2 (a) and (b) The rhombic dodecahedron (a) is a very common form of garnet. It is so often found that it is known as the 'garnetohedron'. Another common form is the trapezohedron (b), which has twenty-four faces, compared to the dodecahedron at twelve faces

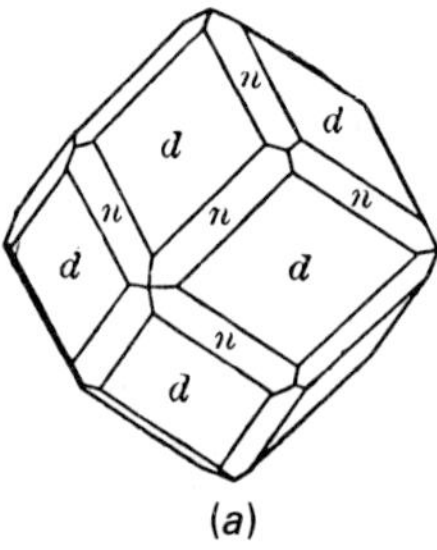

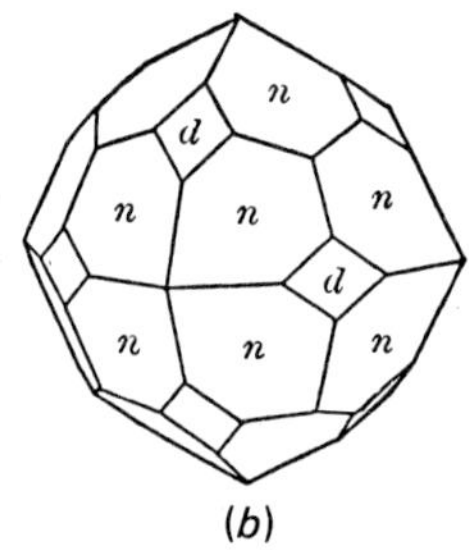

Figure 2.3 (a) and (b) Common forms illustrating combined forms of the simple dodecahedron and the trapezohedron (from Dana, 1896)

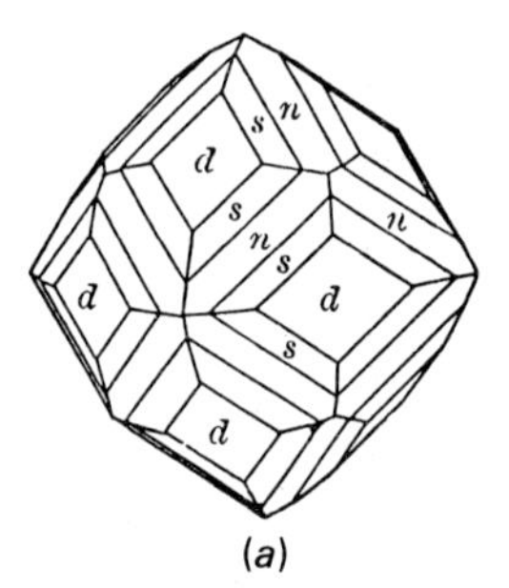

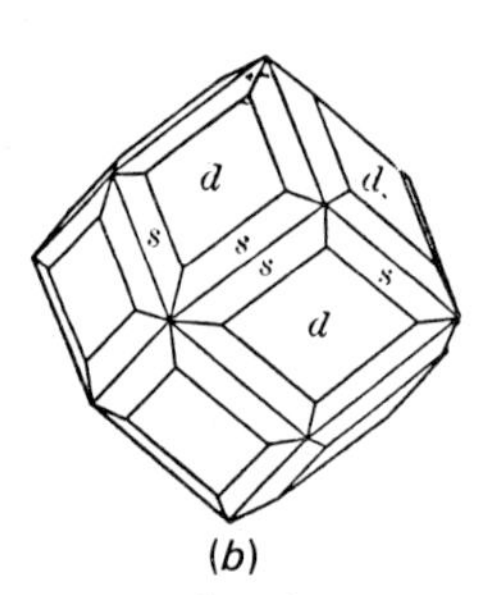

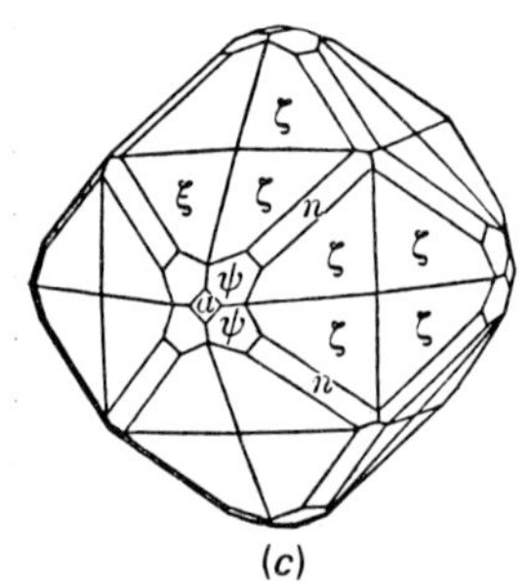

Figure 2.4 (a) to (c) Complex forms are rare, but they can be found developing beyond the simple dodecahedron and the trapezohedron, with additional faces. (c) is a very rare form found in Mill Rock, New Haven, Conn (from Dana, 1896)

forms are complex variations of the two basic forms with additional faces (*Figure 2.4a,b,c*).

Knowledge of the crystal forms of the basic shapes can assist in the recognition of rough specimens, particularly when the dodecahedron or trapezohedron faces are encountered. They are highly characteristic forms for garnet crystals.

Physical and optical properties

The hardness of garnets varies with the species, but it is generally reported to be between 6½ to 7½ on Mohs' scale. Early reports indicated a hardness of up to 8 (Bauer, 1905), but Dana's limit of 7½ is now accepted (Dana, 1896). In a recent study by the gemmologists of the University of Barcelona's school of gemmology, a non-destructive hardness test was used to analyze gem garnets (Arbunies *et al.*, 1975).

The investigation utilized the Vickers diamond pyramid attached to a microsclerometer model MHP of Carl Zeiss. The technique provided Vickers hardness units which were then transformed into Moh's units using mathematical calculations. The resulting data revealed that fifty-one almandine garnets (pyralspite) from Ceylon and four from Spain ranged in hardness between 6.81 to 7.48 on Mohs' scale. Light green grossulars from Italy ranged from 6.66 to 6.98 for thirteen specimens. Forty Canadian hessonites varied from 6.46 to 6.90. This study revealed variations of hardness within species, as well as some unusually low hardness properties within the species studied.

The garnets which have exceeded quartz in hardness (7) have been widely used as abrasives. Such garnets are crushed and graded into grit sizes to be used in a wide variety of abrasive materials. Its popularity in industry is due partly to its hardness and partly to its wide distribution and massive deposits.

When garnet occurs as a massive cryptocrystalline form it can be very tough. But some crystalline garnets are brittle and even friable, particularly the granular massive type, but also some crystalline types. The cleavage is imperfect, but sometimes it can be distinct along the 'd' faces (Dana, 1896 , p. 438). The garnet fractures from sub-conchoidal to uneven, and its lustre varies, in most cases, from vitreous to resinous.

The opacity of the garnets varies, depending on species, but it usually ranges from transparent to opaque among the gem varieties. Translucent stones are rarely used as gemstones, except as cabochons. Faceted gemstones require transparent crystals to display their unique qualities. Inclusions in the garnets vary, depending on the species; consequently, they will be explored in the chapters on species.

Since garnets are isotropic, they form in the same manner as diamond and spinel and are singly refractive. This feature is very useful in separating garnets from rubies, since the latter are doubly refractive. Refractive indices for garnets vary with the species, but they are generally high, ranging from about 1.73 to 1.89. With such high indices, the garnets possess a potential for high brilliance, when properly cut and polished.

Specific gravity, or the relative density of the garnets, also varies with the garnet chemistry, the values usually ranging from 3.40 to 4.30. Most garnets are more dense than diamond at 3.52, but less dense than zircon at 4.70. The specific gravity test, especially when coupled with refractive index and absorption spectrum tests, is quite useful for garnet separations, especially within the garnet species. However, specific gravity must be used with caution, especially with mineral specimens which may be highly included with material of a different density than the host garnet. In gem quality specimens, on the other hand, specific gravity figures will be much more reliable than mineralogical samples, provided that the inclusions are not massive in number or size.

Dispersion for all the garnets is relatively low, with one prominent exception (andradite). Dispersion figures are generally between 0.020 and 0.028, compared to 0.044 for diamond. Consequently, garnets do not rely on their spectral fire for their beauty. The one exception is gem andradite, which produces dispersion measurements of 0.057. In the near colourless andradites, spectral fire can play a role in providing a lovely display of colours in the stone (however, the colour of the gemstone masks the effect to a great extent). Dispersion measurements are not commonly taken by gemmologists, since they are not necessary to identify the specimen. But they are included for comparative purposes with other gemstones, and to explain the fascinating phenomenon that they create in the gem andradites.

Crystalline garnets are impervious to acids, except hydrofluoric acid. Fusion alters garnet properties, however, and thereafter, they no longer resist the actions of acids. Most garnets fuse rather easily at 3 on the fusability scale; pyrope is a little more resistant at 3½ to 4; uvarovite, at 6, is considered infusible with the blow pipe (Miers, 1902, p. 503).

It has long been known that garnets react to the magnet. This condition, called ferromagnetism, is found in those garnets containing iron (almandine, almandine-pyrope intermediates, almandine-sepssartites, andradites, etc.). Indeed, some gemmologists have created tests to estimate the strength of the iron content by measuring the oscillations made by the gemstone when it is suspended by a thread and exposed to a magnet (Anderson, 1959; Trumper, 1962). This test might be usefully applied to the sometimes difficult separations between almandine and spessartite.

Another unusual property exhibited by garnet is its ability to generate electricity by rubbing or heating. This characteristic was reported by Pliny (37:29), also by Bauer (1905, p. 349), and more recently by Novak and Gibbs (1970, p. 792). This attribute is usually associated with tourmaline, and has caused confusion when ancient texts described the feature and left readers to wonder if the reference was meant to be tourmaline rather than garnet. However, since there are no surviving samples of tourmaline from the ancient world, and since this property is found in garnet as well, then garnet must be meant.

Garnet colours

Garnets occur in all colours except blue. The orange-red, red and violet hues are most commonly associated with both pyrope and almandine species; grossular ranges in a wide variety of colours, from the rare colourless, to orange, yellow, yellowish-green and green; green is also found in the demantoid garnets and the uvarovites, as well as knorringite (blue-green); spessartite occurs in a distinct array of orange to reddish-orange hues, sometimes with a pink overtone, reminiscent of the beautiful padparadscha sapphire (see *Figure 2.5a*).

Gemstone colours, like colours in general, are often difficult to convey. Writers commonly describe them in terms of special names which have only limited meanings within certain contexts. Occasionally, they may refer to animals, plants or other natural phenomena, as in this ancient colour description of gem garnet:

'Of garnets, those which are coloured like the black spot in the Gunja, like honey, the stalk of the lotus, the musk deer, fire, or the plantain-tree, are first rate. A garnet which is coloured like the conch, the lotus, the black bee, or the sun, and which is strung on a thread, is sound and auspicious, and heralds good fortune. A garnet which is coloured like the crow,

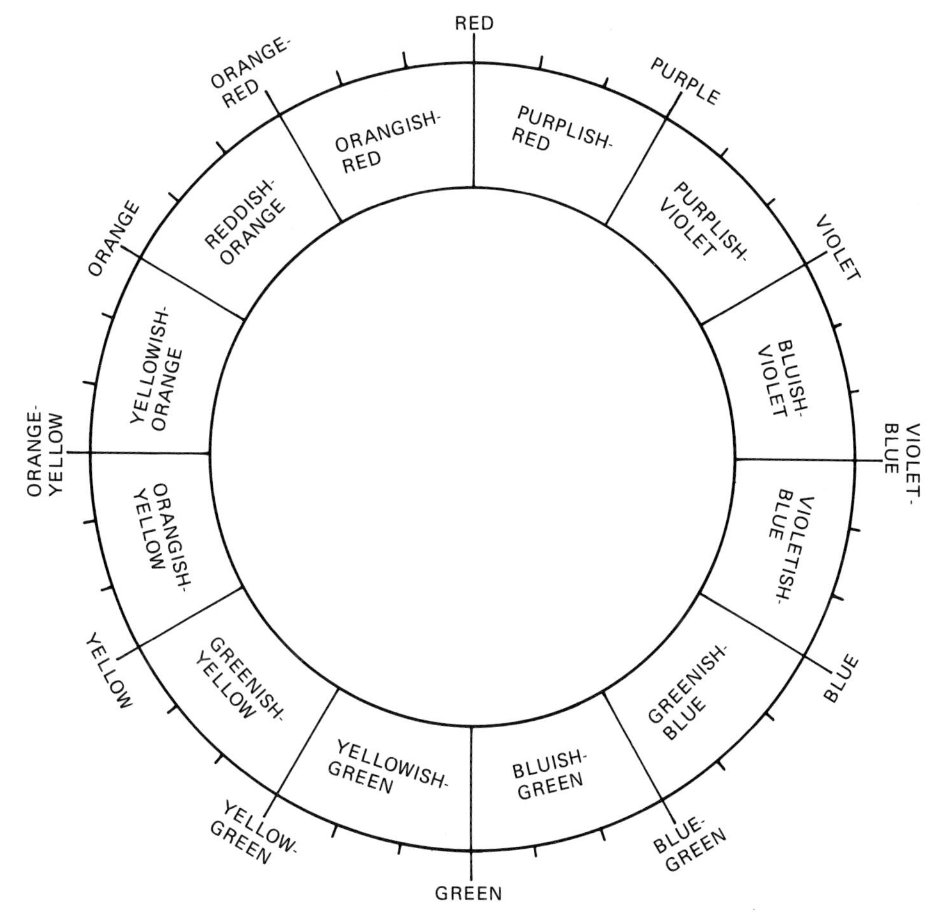

Figure 2.5 (a) Hue positions on a two-dimensional colour wheel. Combined with intensity and tonal positions, this system is perhaps the most precise method for describing gemstone colours

the horse, the ass, the jackel, the bull, or the blood-stained beak of the vulture holding a piece of flesh, bring on death; and the authorities advise us to shun it'. (Ancient Arabic description, translated by Tagore, *Mani-Mali,* 1879)

However, there are other colour names that are restricted to certain nationalities, or are abstracted from reality to such an extent that they are not widely known. Puce is a good example; it is supposed to convey a brownish-purple and is derived from the French word for flea. Another is kelly, designating a type of green. Carmine, magenta, cobalt, azure and mauve also designate colours used in fashion or industry that convey a vague, or restricted meaning.

In scientific colour systems used widely in science and industry around the world, colours are described in terms of a colour-universe, divided into hue, tone and intensity. Hue is defined as the colour of the sample, as it might appear on a colour wheel. The ISCC (Inter Society Color Council) is an American group of colour experts who have assigned colour names to their book of colour chips. Each chip is scientifically created in connection with the National Bureau of Standards and is named in accordance with the colour space in the ISCC colour universe. Another system is the American Munsell system, which also describes a universe of colour divided into hue, chroma (intensity) and value (tone). Hue pages are designated in colour-wheel terms and the system is appearance related and can accommodate millions of colours. These systems are used to convey precisely a colour description by describing the hue-mix positions of the colour wheel. In practice, gemstone colours can be much better defined using such nomenclature, rather than the methods described earlier (see *Figure 2.5b*), detailing the ISCC-NBS colour names in a three dimensional colour space.

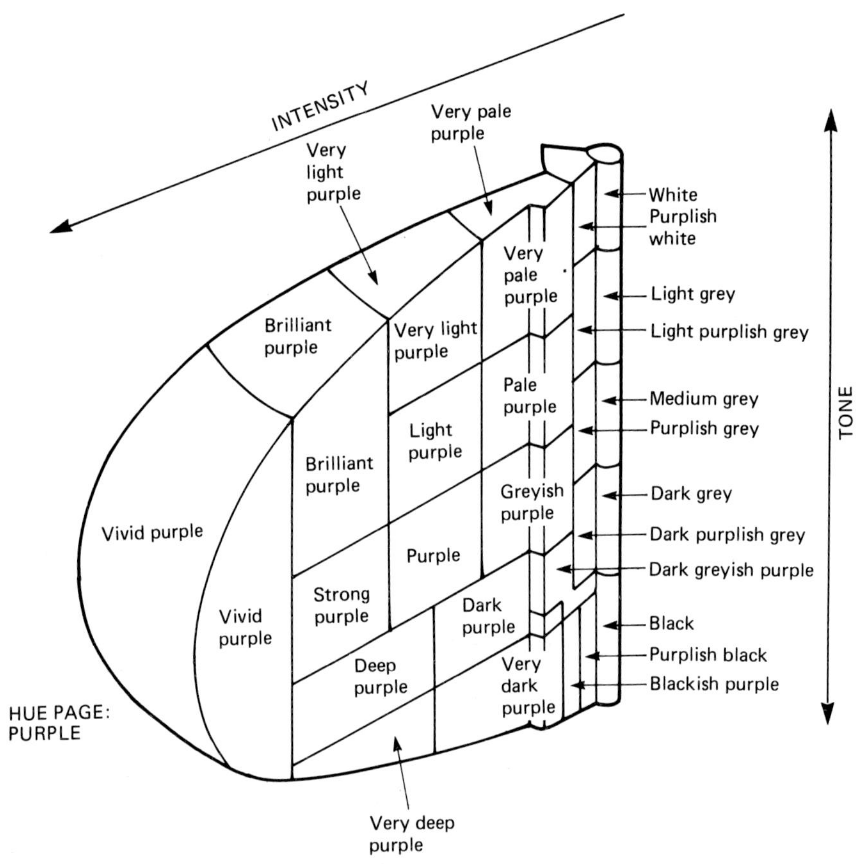

Figure 2.5 (b) The ISCC-NBC colour names for purple within a three-dimensional colour space universe (Billmeyer and Saltzman, 1981)

There is a current trend in using these scientific systems (and several others) to better describe gemstone colours. Jobbins *et al.* (1978) completed a study of 204 garnets from Tanzania in which the Munsell colour terminology was applied. In another study by the author (see Appendix 1), an analysis of Malagasy and Orissa (India) garnets used CIE nomenclature along with two other visual colour systems in a comparative analysis. Other gemmological studies are also encountered using various scientific colour systems in an effort to define the gem colours with greater precision.

In addition to the garnet hues, described above, the tones and intensities also vary considerably. In gemstones, ideal tones are found in the mid-range. Hues that are too dark or too light offer either a hidden beauty or an absence of beauty, respectively, both conditions detrimental to the gem's appearance. Dark tones are the lesser problem of the two, since the colour may be rich and beautiful, but requiring a strong light to make it visible. Both almandine and pyrope garnets have traditionally very dark tones, while the grossular exhibits problems with excessively pale tones. Exceptions to the general rule, however, produce very striking gemstones which are appreciated and command the highest prices within the species (depending on the hue position and richness of the intensity, of course).

Intensities also vary in the garnets, even within the same species. Intensity represents the purity of the hue. Stated another way, intensity is the amount of grey in a hue or hue-mix. If an artist prepares a pool of pure hue (or hue-mix), and begins to dilute the hue by adding drops of grey paint, a certain amount of 'dulling' occurs. If a large amount of grey is poured into the hue or hue-mix, then the intensity becomes dull and lacks brightness and richness. This is an important dimension of colour, for the most beautiful gems exhibit high degrees of intensity. A scale for this colour dimension can range from dull to weak to moderate to bright to vivid. With some practice, gemstone colours can be placed accurately within this scale of intensity.

Of course, the activity must be conducted with a proper grading light and viewing techniques. Garnets can be rich in hue intensity. The reds of the chrome pyrope can vie with similarly coloured rubies; the richness of the spessartite is often compared to the magnificence of the sunset; and the green of the tsavorite or demantoid can only be excelled by the finest emeralds. Garnet hues can be very intense. But such colours must be sought out and studied before they can be properly appreciated.

The cause of colour in some garnets is due to the presence of trace elements of metal oxides in the chemical mix, and in others by metal oxides inherent in the stone's chemistry. If the colour of the gemstone is due to the metal oxide in the trace chemistry of the species, it is said to be an allochromatic gem ('allo' = other; 'chromatic' = colour: the colour is derived from a foreign substance). On the other hand, if the colour of the gem is due to an oxide that is inherent in the gemstone's own bulk chemistry, then it is known as an idiochromatic gem ('idio' = from within one's self).

Traditionally, the idiochromatic garnets are almandine, spessartite and uvarovite (see *Table 2.3*). The allochromatic garnets are pyrope, coloured by iron or chromium, and grossular, coloured by manganese, iron, chromium and/or vanadium. Pure allochromatic gems are colourless, and they are known to exist in the grossular garnets, but not the pyropes. Andradites could be either allochromatic or idiochromatic, depending on the variety. Demantoid is coloured by chromium, which is not part of its bulk chemistry; topazolite (the yellow andradite) is coloured by the iron of its bulk chemistry, therefore it is an idiochromatic gemstone.

TABLE 2.3. Idiochromatic and allochromatic garnets

Idiochromatic: 'Self-coloured'

The colour of the stone results from an essential element in the stone's bulk chemistry formula. Thus, idiochromatic gems are limited in colour to the range produced by the particular element which causes the colouration.

Example:		
	Turquoise:	Blue, green, or blue-green Colour due to Cu
	Peridot:	Green, yellow, or green-yellow Colour due to Fe

Idiochromatic garnets:

Almandine:	Red, purplish-red, purple, or orange-red Colour due to Fe
Spessartite:	Orange, reddish-orange, yellow orange Colour due to Mn
Uvarovite:	Green, yellowish-green Colour due to Cr
Andradite (Topazolite):	Yellow, greenish-yellow Colour due to Fe

Allochromatic: 'Other-coloured'

Colourless if chemically pure. The colour is produced by foreign elements substituting for essential elements of the bulk chemistry, usually in small amounts.

Example:		
	Corundum:	Al_2O_3: if pure = colourless Al_2O_3 + Cr = Ruby Al_2O_3 + Fe + Ti = Blue sapphire

Allochromatic garnets

Pyrope:	$Mg_3Al_2(SiO_4)_3$ + Cr = Intense red colour $Mg_3Al_2(SiO_4)_3$ + Fe = Orange-red, violet-red
Grossular:	$Ca_3Al_2(SiO_4)_3$ = Colourless $Ca_3Al_2(SiO_4)_3$ + Mn = Orange to yellow orange $Ca_3Al_2(SiO_4)_3$ + Mn + Fe = Reddish-orange to brownish-orange
Andradite (Demantoid):	$Ca_3Fe_2(SiO_4)_3$ + Cr = Green, yellowish-green

The colours of the garnets should not be used as a means of garnet identification. In some cases, the garnet colour can be used as an indication of a garnet species or variety, but the stones should always be checked with instruments to determine identity.

Garnet classification

An important reason for knowing the properties and colours of the garnet species and varieties is to properly classify unknown garnet specimens. Many studies have concentrated on these characteristics with the main purpose to define carefully (or redefine) garnet parameters. In the important study by Skinner (1956), pure end-member properties were reported from synthetic garnets produced in pyrope, almandine, spessartite, grossular and andradite. Earlier studies of natural garnets provided little information on the pure end-member species, for most species did not occur in pure end-member specimens.

A relatively early study reported a definite relationship between the refractive index, specific gravity and the chemical composition of the garnets (Ford, 1915). Furthermore, it was suggested that if the refractive index and the specific gravity of the garnet were known, and knowing from quantitative tests the predominant molecules present, that it should be possible to estimate within a reasonable degree of error the chemical composition of the specimen. This study was supplemented and refined later by Stockwell (1927), Fleischer (1937), Wright (1938), Levin (1950) and others.

The literature was searched for published samples of garnets with known compositions and properties. However, the approach to this effort was not conducive to gemmological needs, for the mineralogists used tests which were not applicable to gemstones (X-ray analysis and a blowpipe test for manganese, etc.). The gemmologist, on the other hand, must utilize non-destructive tests (refractive index, specific gravity, absorption spectrum, and sometimes inclusion studies) in order to classify the garnets into species and varieties.

Anderson (1959) reacted to the issue and clarified the problems for the gemmologist. The conflict between pyrope and almandine could be resolved by adopting RI and SG limits separating the two species, and by creating an intermediate category, which he called 'pyrandine'. Also, the complex question between the sometimes confusing spessartite and almandine could be resolved by the study of the absorption bands of those garnets. Furthermore, since the 'ugrandite' series did not exhibit the same identity crises, those species could be resolved rather easily.

This landmark study seemed to fill a need of the times, for most known pyropes of the period would fit comfortably within the 1.73–1.75 limits suggested by Anderson (especially the pyropes of Bohemia, Africa and Arizona). Moreover, intermediate garnets (rhodolite) were known to fit within the gap of 1.75–1.78, while known almandines could be found above 1.78. However, what was overlooked was the fact that almandines graded right into the intermediate category, while other sources of 'rhodolite' graded into pyrope. Moreover, the discovery of the 'Malaya' garnets added to the confusion, for they were found from pyrope to almandine, including many samples in the intermediate category.

Manson and Stockton's studies in the garnet group seemed to suggest the need for an intermediate category, as many stones in their study fell into a definable intermediate position, irrespective of colour or source (Manson and Stockton, 1981; Manson and Stockton, 1982). Even Webster's newest edition, edited by B.W. Anderson before his death still described the intermediate category, although Anderson's original name 'pyrandine' was replaced by 'pyrope-almandine' (Webster, 1983). So many gemstones can be found in this intermediate category, that the recommendation cannot be ignored: either the point between pyrope and almandine is to be decided upon in an arbitrary manner, eradicating the intermediate category, or the intermediate category is to be retained with arbitrary limits to either side of the classification.

Another more recent proposal for garnet reclassification has been one made by Dr. Hanneman (1983), after reading the studies by Manson and Stockton. Hanneman suggested alternative limits for the three categories and proposed a new garnet variety name: 'chrome-pyrope' for those pyropes rich in chromium. Furthermore, he suggested the removal of SG limits within the three categories, using only the refractive index and the absorption spectrum for proper classification.

Certainly the classification issue with regard to the garnet family is far from over. It does seem evident, however, that RI, SG and the absorption spectrum will continue to be necessary for precise garnet classification. Changes within the nomenclature may be difficult to justify, unless there is substantial evidence arising from the study of many gem garnets from known worldwide sources.

In the studies just cited, it seems obvious that there is a need for an intermediate category, whether or not the name is suitable for the gemstones. If the category is to be called 'pyrope-almandine' as suggested by Manson and Stockton, 'pyralspite' as proposed by Bank (Manson and Stockton, 1982), or 'pyrandine' (Anderson, 1947), or some other name, it is certainly unsettled at this time.

Garnet identification

For the identification of the garnet family of gemstones, gemmologists concentrate on the RI, SG and the absorption spectrum of the pyralspite series (*see Table 2.4*).

TABLE 2.4. Chief identification characteristics of the major gem garnets

Pyrope identification
Refractive index: 1.73–1.75 (Key to ID)
Specific gravity: 3.65–3.80 (Key to ID, when coupled with RI)
Absorption spectrum:
Iron-rich: Characteristic spectra
Chromium-rich: Characteristic spectra (key to ID)

Almandine identification
Refractive index: 1.78–1.82 (key to ID)
Specific gravity: 3.95–4.30
Absorption spectrum: Characteristic iron spectra (key to ID)

Intermediate pyrope-almandine identification
Refractive index: 1.75–1.78 (key to ID)
Specific gravity: 3.80–3.95

Spessartite identification
Refractive index: 1.79–1.81
Specific gravity: 4.12–4.20
Absorption spectrum: Characteristic Mn bands (key to ID)
Colour: May overlap almandine, generally orange, however

Grossular identification
Refractive index: 1.73–1.76 (overlaps pyrope)
Specific gravity: 3.40–3.78 (key to ID)
Colour: Many colours characteristic to species (key to ID)

Andradite identification
Refractive index: 1.880–1.888 (key to ID if measurable)
Specific gravity: 3.77–3.88 (key to ID)
Colour: If green, demantoid, if yellow to greenish-yellow, topazolite
Inclusions: 'Horsetail' inclusions (key to ID if observed)
Dispersion: Spectral colours usually visible in lightly coloured specimens

For grossular garnets, the absorption spectrum is not characteristic, but colour, RI and SG properties are used with definitive results. In the andradite species, colour, RI, SG and characteristic inclusions ('horsetail') are used for identification. The varieties of demantoid and topazolite are separated on the basis of colour: green for the former and yellow to greenish-yellow for the latter.

The separation of garnet from ruby is accomplished by means of the dichroscope (ruby will exhibit two colours in most directions, while garnet will only reveal one colour in any direction).

Separating garnet from spinel is accomplished with the spectroscope. Spinel reveals characteristic spectra, quite differnt from garnets. The separation of garnet from other substances and gemstones is easily accomplished by means of RI, SG and absorption spectrum studies.

Garnet treatments

Garnet is one of the few gemstones that does not react to known treatments. Consequently, sellers of garnet need not disclose any man-made alteration to their customers.

Although synthetic garnets have been created in the laboratory, because of the cost involved and the selling price of most garnets, they have not been produced commercially.

Bibliography

ANDERSON, B. W., 'Pyrandine – a new name for an old garnet', *The Journal of Gemmology* (April 1947)

ANDERSON, B. W., 'Properties and classification of individual garnets', *The Journal of Gemmology,* VII, No. 1 (January 1959)

ARBUNIES-ANDREU, M., BOSCH-FIGUEROA, J. M., FONT-ALTABA, M. and TRAVERIA-CROS, A., 'Physical and optical properties of garnets of gem quality', *Fortschritte Mineralogie,* **52** (1975)

BAUER, MAX, *Precious Stones,* translated by L. J. Spencer, Rutland, Vermont and Tokyo, Japan, 1970 (1905)

DANA, EDWARD SALISBURY, *The System of Mineralogy,* 6th edition, London (1896)

DEER, N. A., HOWIE, R. A. and ZUSSMAN, J., *Rock-forming Minerals, Ortho- and Ring Silicates,* Vol. I, London (1962)

FLEISCHER, M. H., 'The relation between chemical composition and physical properties in the garnet group', *American Mineralogist,* **22** (1937)

FORD, W. E., 'A study of the relations existing between the chemical, optical and other properties of the garnet group', *American Journal of Science,* 4th Series (1915)

HANNEMAN, W. WILLIAM, 'A new classification for red-to-violet garnets', *Gems & Gemology* (Spring, 1983)

JOBBINS, E. A., SAUL, J. M., STATHAM, PATRICIA M. and YOUNG, B. R., 'Studies of a gem garnet suite from the Umba River, Tanzania', *Journal of Gemmology,* **XVI,** 3 (July 1978)

LEVIN, S. B., 'The physical analysis of polycomponent garnet', *American Mineralogist,* **35** (1950)

MANSON, D. VINCENT and STOCKTON, CAROL M., 'Gem garnets in the red-to-violet colour range', *Gems & Gemology* (Winter 1981)

MANSON, D. VINCENT and STOCKTON, CAROL M., 'Gem garnets: the orange to red-orange color range', *International Gemological Symposium Proceedings,* ed. by Dianne M. Eash (1982)

MIERS, HENRY A., *Mineralogy,* London (1902)

NOVAK, G. A. and GIBBS, G. V., 'The crystal chemistry of the silicate garnets', *American Mineralogist,* **56** (1970)

PLINY, *Natural History,* trans. by D. E. Eichholz, Loeb Classical Library Edition (1971)

RICKWOOD, P. C., 'On recasting analyses of garnet into end-member molecules', *Contributions to Mineralogy and Petrology,* **18** (1968)

SKINNER, B. J., 'Physical properties of end-members of the garnet group', *American Mineralogist,* **41** (1956)

SRIRAMADAS, A., 'Diagrams for the correlation of unit cell edges and refractive indices with the chemical composition of garnets', *American Mineralogist,* **42** (1957)

STOCKWELL, C. H., 'An X-ray study of the garnet group', *American Mineralogist,* **12** (1927)

STOCKTON, CAROL M. and MANSON, D. VINCE, 'Gem andradite garnets', *Gems & Gemology* (Winter 1983)

TAGORE, SOURINDRO MOHUN, *Mani-Málá, or A Treatise on Gems,* Part 1, Calcutta (1879)

TISDALL, F. S. H., 'Tests on Madagascar Garnets', *The Gemmologist* (June 1962)

TRUMPER, L. C., 'Observations on garnet', *The Journal of Gemmology* (October 1962)

WEBSTER, ROBERT, *Gems,* 4th edition, revised by B. W. Anderson, London (1983)

WINCHELL, A. N. and WINCHELL, H., *Elements of Optical Mineralogy,* 4th edition, New York (1933)

WINCHELL, HORACE, 'The composition and physical properties of garnet', *The American Mineralogist,* **43** (May-June 1958)

WRIGHT, W. I., 'The composition and occurrence of garnets', *American Mineralogist,* **23** (1938)

Chapter 3

Pyrope garnet

Although pyrope garnet has had several centuries of unprecedented popularity, today the gem is ubiquitous and not highly valued. Reasons for this change of opinion are discovered when analyzing both the supply and the demand factors behind the popularity.

Although the supply of pyropes in the early 19th century seemed vast in the large garnet fields of Bohemia, these mines had been working, perhaps continuously, since the 16th century, and probably long before. As a major source of almost exclusively pyrope garnet, were there in 1800 adequate supplies available from Bohemia to last through a high-demand period that might last for decades? However, the supply side did manage to hold out through the 19th century.

But the demand was much more volatile. The industrial revolution and the advent of machine-made jewellery became coupled with the lavish display of gems and jewellery inspired during the Victorian period (see Flower, 1951, for a full description of the times and fashion). Gemstones were in vogue and a huge demand was created. The peak of that demand can be seen reflected by Kunz in 1892. But the end was already in the making; for fashion was even then moving in more austere and new directions.

The saturation of the marketplace with cheaply-made Bohemian jewellery made an indelible mark in the minds of the public, both in America and in England. The memory still lingers on, and pyrope garnet has fallen with the fashion. Only the relics remain, either on display in solitary museum cases, or in long ignored jewellery boxes of countless grandmothers on several continents. Yet, despite this public reaction, there are today avid collectors, hobbyists and gem connoisseurs who still search for rare examples of the fiery, red pyrope.

Pyrope colours

The term 'pyrope' is derived from the Greek words, πup, meaning 'fire', and ὄπτομαι, meaning 'I see'. The fiery qualities of the garnet are especially seen in the modern species of pyrope. Varying in colour from slightly orangy-red to red, and on to violet-red as the iron from the almandine alters the hue position, pyrope can exhibit very striking, bright and fiery hues. Moreover, the viewing position of ancient gem enthusiasts was accomplished under transmitted light with rough material or cabochon-cut gemstones. Under such conditions, the fiery qualities are evident, but not to the same degree that colour is displayed in modern faceted gemstones.

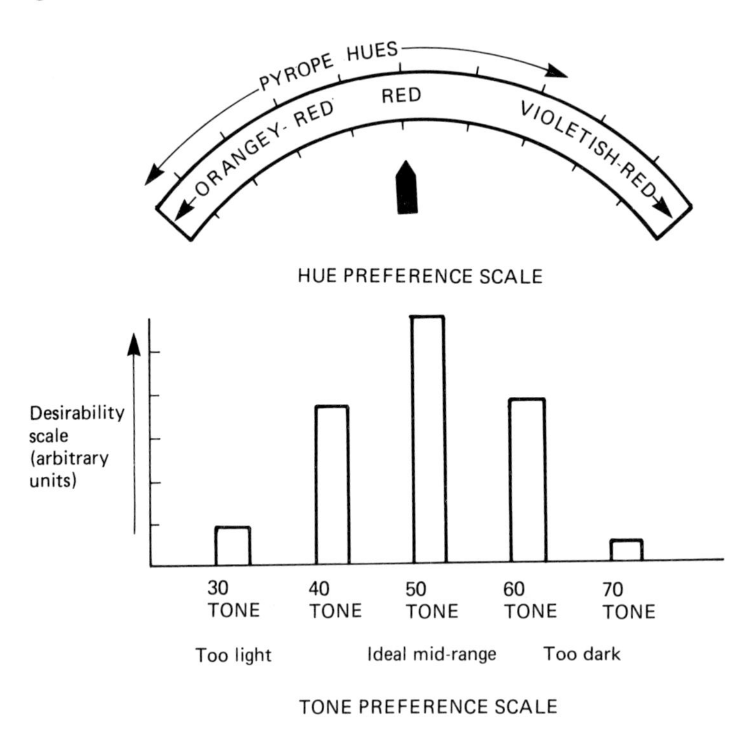

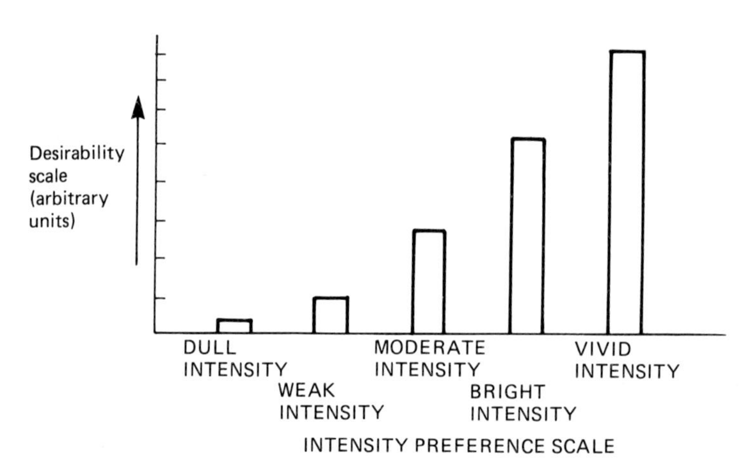

Figure 3.1 General market preferences for hues, tones and intensities of pyrope

Modern cutters have long noticed a considerable change of colour when comparing rough specimens to finished gems. The faceted stone relies on internally reflected light which produces a more intense hue, often accompanied by a slightly different hue position than the colour seen under transmitted light. The overall appearance of the gemstone is substantially improved with faceting.

Unfortunately, most pyropes also exhibit a dark tone. As in all dark-toned gemstones, a stronger light is necessary to bring out the richness of the 'fiery' colour. When pyropes are viewed under strong lighting, particularly well-balanced grading lamps, the beauty of the gem is quite evident. Collections of pyropes with varying hues can be studied under such lights to provide hours of stimulating and instructive pleasure. However, not all pyropes display dark tones. Those pyropes in the ideal mid-tonal range are often sought after by collectors. Their rarity and more visible beauty make them more desirable specimens, often rivalling some rubies in their appearance.

Intensities of pyrope hues tend to be high, especially in the reds with moderate tones. However, as the hue position moves in either direction, whether into the orangy-reds or into the violet-reds, the intensity of the pyrope usually, but not always, weakens. Moreover, the few light-toned pyropes also exhibit only moderate to dull intensities (see *Figure 3.1* for illustrations of pyrope colour preferences in the gem marketplace).

Pyrope classification

Although pyrope is one of the most common of the modern garnets, it was not classified as a species until the last century, when chemical testing became routine. However, there was a reference to 'Bohemian' carbuncle by Agricola (*De Natura Fossilium*) in 1546 and de Boodt (*Gemmarum et lapidum*) in 1609; this was undoubtedly pyrope, for there are few almandines or other garnets from that source.

Thomas Nicols (1652) described 'pyrope' as a synonym for carbuncle (p. 56). Consequently, the term was not yet associated with garnet, nor with the Bohemian gemstones which he declared were garnets (p. 62).

Interestingly, the Bohemian pyropes were tested in this early period with a blowpipe in order to determine whether or not they were true garnets, for the Bohemian gems 'will endure the fire without any loss of colour, and with little or no harm . . .' (p. 62). More recent studies with the blowpipe suggest that pyrope is somewhat more difficult to fuse than most other garnets (see Miers, 1902, p. 503), perhaps confirming Nicol's test.

Pyrope chemistry: $Mg_3 Al_2 (Si O_4)_3$

Modern chemical analyses has established that pyrope is a magnesium aluminum silicate. In order to compare compositions of natural pyrope, a number of samples are presented in *Table 3.1*. It can be seen that almandine is commonly mixed with pyrope, as the molecule percentages vary from a low of 11.3% to a high of 30.99%.

TABLE 3.1. Pyrope properties and end-member calculations from chemical analyses of natural pyrope reported in the literature

Sample	1	2	3	4	5	6	7	8	9	10	11	12	13	14
RI	1.739	1.739	1.739	1.741	1.742	1.743	1.746	1.749	1.750	1.750	1.752	1.753	1.754	1.761
SG	3.745	–	3.668	–	3.715	3.712	3.690	3.683	3.782	–	3.793	3.82	3.796	3.786
Molecular end-member composition (calculated %)														
Pyrope	73.0	55	72.7	55.30	61.66	73.7	74.2	73.3	60.7	54.78	50.5	60.7	54.9	50.7
Almandine	14.6	20	11.3	28.83	24.04	13.9	13.6	15.0	26.3	30.99	26.9	33.7	30.3	37.3
Spessartite	8.3	–	2.0	0.75	1.15	0.6	0.6	0.5	0.7	0.88	0.8	2.3	0.6	4.7
Grossular	2.4	25	5.7	5.96	5.31	3.1	1.3	4.6	8.3	8.61	20.9	3.1	10.1	6.8
Andradite	1.6	–	6.4	–	–	3.3	2.1	5.3	3.4	–	0.9	0.2	2.7	–
Uvarovite	0.1	–	2.0	9.16	7.48	5.4	8.2	1.3	0.6	4.74	–	0.2	–	0.3
													Other:	0.1

Sources:
1. Pyrope, alluvial deposit, Umba, Tanzania; Jobbins, *et al.* (1978)
2. Pyrope from eclogite, Bavaria; Wright (1938)
3. Pyrope from eclogite, So. Moravia; Deer, Howie and Zussman (1962); recalculated by Rickwood (1968)
4. Pyrope from Kimberley, So. Africa; Ford (1915)
5. Pyrope from the Colorado River, Arizona; Ford (1915)
6. Chrome pyrope: Lherzolite nodule, pipe at Farm Louwrencia, nr. Gibeon, S.W. Africa; Nixon, *et al.* (1963)
7. Chrome pyrope: heavy mineral washings (concentrate), Kao pipe, Basutoland, So. Africa; Nixon, *et al.* (1963)
8. Pyrope: discrete nodule, Thaba Putsoa pipe, Basutoland, So. Africa; Nixon, *et al.* (1963)
9. Pyrope from Rodhaugen, Sondmore, Norway; Fleischer (1937); recalculated by Rickwood (1968)
10. Pyrope from Kimberley, So. Africa; Ford (1915)
11. Pyrope from zoisite eclogite, Silberbach, Bavarian Fichtelegebirge, Germany; Deer, Howie and Zussman (1962) recalculated by Rickwood (1968)
12. Pyrope from Umba, Tanzania; Jobbins, *et al.* (1978)
13. Pyrope from pyroxene-garnet rock, Salem district, Madras, India; Deer, Howie and Zussman (1962), recalculated by Rickwood (1968)
14. Pyrope from Umba, Tanzania; Jobbins, *et al.* (1978)

Since the mixture is quite soluble, any proportion of the two molecules can be expected to occur in natural samples. The grossular component also varies from 1.3% to 25% in the samples, possibly suggesting a similar solubility relationship, despite early views that the two series 'pyralspite' and 'ugrandite' rarely mixed (Winchell and Winchell, 1933). The solubility between pyrope and spessartite does not appear to be high in these samples, since the spessartite component only reaches 8.3%. Pyropes high in chromium (Cr_2O_3), will be reflected in high uvarovite percentages. Those samples with over 4% uvarovite (molecular percentage) could be considered 'chrome' pyropes (Nixon *et al.*, 1963).

Colouration for pyropes is usually attributed to the iron of the almandine or the chromium of the uvarovite (or other garnet molecule). Other minor trace elements may also be involved because the colouration in pyrope can be quite varied. The bright red of the pyrope is caused by the chromium content in the sample. The violet or orange-red colours are commonly thought to be due to the ferrous iron of the almandine molecule. Variations between violet-red and orange-red colours in pyrope (as well as in almandine) perhaps require further studies into the relationships between colour and chemistry in order to accurately establish the exact cause of the colour.

Optical and physical properties

It is difficult to establish limits for the refractive index of pyrope. In the pure state, pyrope does not seem to exist; however, it has been synthesized in its pure state, revealing an RI of 1.714 (Skinner, 1956). In nature, the refractive index can be found from about 1.730 and upward. Arbitrarily, a point is selected in the solid solution series between pyrope and almandine in order to define the pyrope species. Anderson's recommendations are largely followed in Europe (Hanneman, 1983). He recommended an upper limit of 1.750 and a specific gravity range from 3.65–3.80. Most pyropes from important historical sources (Bohemia, South Africa and Arizona) reportedly fit into this category (Anderson, 1959), but *Table 3.1* reveals some interesting exceptions.

It can be seen from *Table 3.1* that 'pyrope' as defined by chemical analysis can reach RI property values as high as 1.761 (sample No. 14). Conceivably, there could be natural samples which proceed higher than the 1.76, depending upon the amount of the ferrous iron in the stone. This example only serves to reinforce the arbitrary nature of Anderson and Webster's traditional limits for pyrope, established at 1.750 (Webster, 1983).

Absorption spectra of pyrope

Iron and chromium have long been recognized as components of pyrope. Both elements exhibit characteristic spectra which can be used to identify the species and to separate iron-rich samples from chromium-rich stones. Because of this potential separation, Hanneman (1983) based his recommendations for 'chrome-pyrope' to be classed as a separate variety (see discussion on general classification of garnets, above).

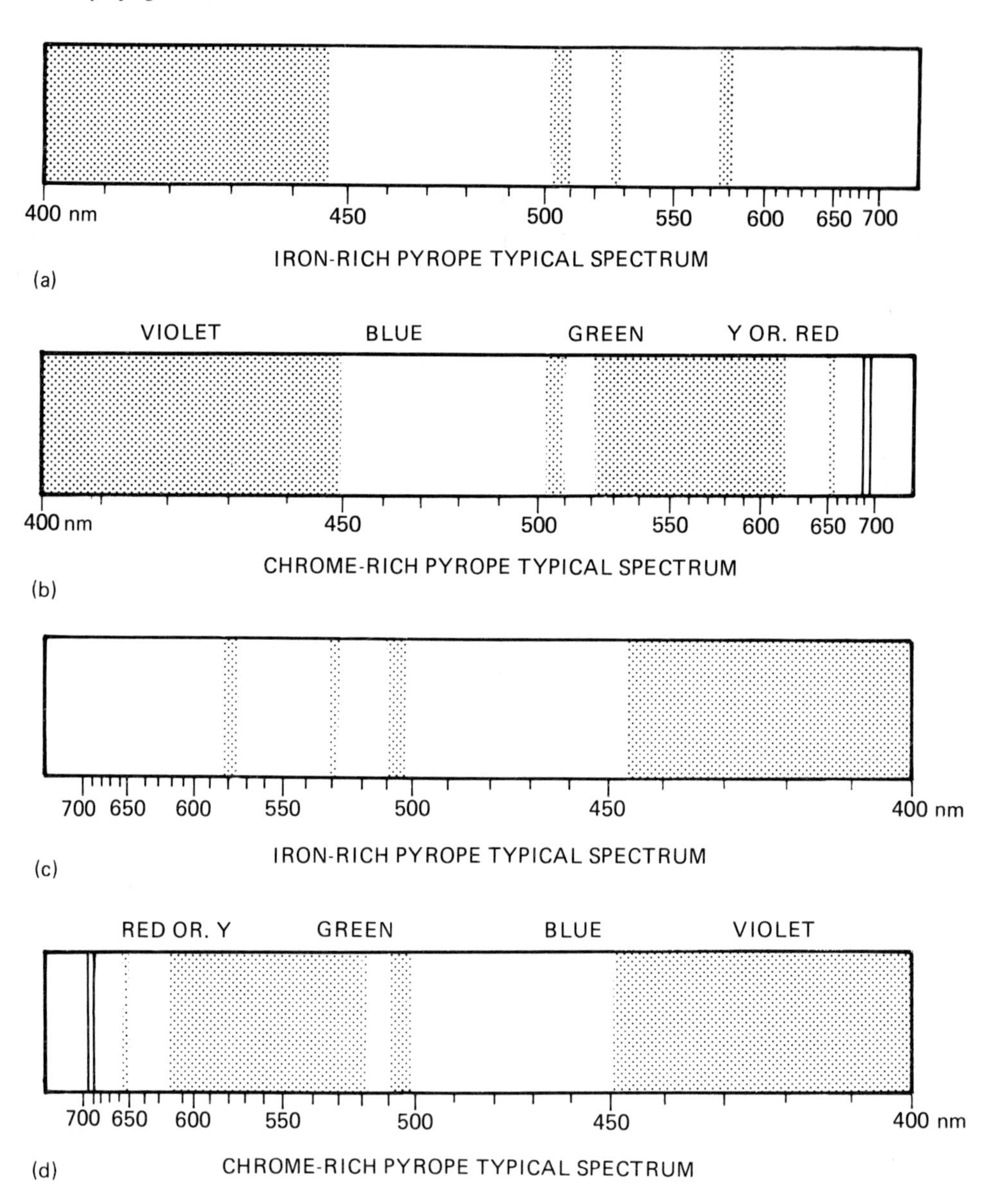

Figure 3.2 Typical spectra of iron-rich and chromium-rich pyrope. (a) and (b) USA; (c) and (d) Europe

Important studies by Anderson and Payne (1954) revealed that most rich, red gem pyropes from the famous sites of Bohemia, Arizona and South Africa contained enough chromium to be identified in the spectroscope. They also acknowledged the existence of 'occasional' pyropes with low properties and no chromium. It is suspected, however, that with the rarity of fine, rich red pyropes from those classical sites and with the more recent abundance of iron-rich pyropes from new sources, the iron-rich stones are more available today than in 1954. Nevertheless, both types of pyrope can be readily identified with the spectroscope.

Since pyrope nearly always contains some iron, the iron bands are revealed in the spectroscope in the classic positions of 505 nm, 527 nm and 575 nm. The bands appear stronger as the iron content is increased from the low-property pyropes to the high-property specimens. However, even in high-property pyropes, the strength of those bands is just in the moderate range, and other typical almandine bands are not visible (Rouse, 1986). If these three bands are seen, then it can be assumed that there is little or no chromium in the chemical mix.

The chromium is quite apparent, even in those chrome pyropes with a considerable iron content. The strong chromium band occurs between 520 nm and 620 nm. Although this diagnostic band covers the iron bands (if iron is present) at 527 nm and 575 nm, yet one single iron band remains at the 505 position. The strength of the 505 nm band will indicate the strength of the iron present, at least in a general way. In addition to the broad chromium band, there are additional chromium lines, which, due to their moderate weakness, may not always be seen in the spectroscope. Moderate chromium lines occur at 650 nm and 671 nm. Additional chromium lines, of moderate strength, occur as a doublet, with fine lines at 685 nm and 687 nm (see *Figure 3.2*).

Pyrope inclusions

Pyrope tends to be freer of inclusions than other garnets. However, as pyropes contain more of the almandine end-member, and are commonly referred to as high-property pyropes (approaching 1.750 in RI and 3.80 in SG), invariably more inclusions are found. Acicular crystals of horneblend were reported in a high property pyrope (RI, 1.750), seemingly randomly placed, but found to be parallel to the dodecahedron faces (Trumper, 1952).

In another study by the same author, ten years later, various inclusions were reported – crystals and short needles, 'short rods', rounded crystals, silk, and dendritic feathers. The 'short rods' were also seen in a more recent study of high property pyropes from Orissa, in addition to two-phase inclusions (Rouse, 1984). Bohemian garnets frequently contain quartz crystals or other crystals (augite) arranged in a circular pattern (Webster, 1983).

Sizes of pyrope

Pyropes tend to be small in size. They have been occasionally reported in large sizes, however. Kaiser Rudolph possessed a pyrope from Bohemia the size of a pigeon's egg (de Boodt, 1609, p. 77); another pyrope of fine quality was found in the size of a hen's egg (Bauer, 1905, p. 358); a third weighing 468.5 carats was reported in the 'Green Vaults' at Dresden, and set in an Order of the Golden Fleece (Bauer, 1905, p. 358). These were exceptional specimens, however.

In the Bohemian mines, only rarely were large stones found. It was reported that for every two tons of garnet mined at Bohemia, one stone occurred of about five carats; and for every 220 pounds, there was one stone of about 2.5 carats; however, one carat stones and below were progressively common (Bauer, 1905, p. 358).

In the two main sources for information on the Bohemian mines (Bauer and Kunz), there is disagreement on the basic measurement of the 'loth'. Kunz (1892) declared the 'loth' to be one-eighth of a troy ounce, while Bauer claimed the loth to be 16.66 grams, somewhat less than half of an avoirdupois ounce (Kunz, 1892, p. 3; Bauer, 1905, p. 358). If Kunz was correct, then stones at their largest average size would be just under one carat (0.97 cts). The Bauer stones in the previous paragraph are calculated according to his estimate of the loth.

In the Arizona pyrope deposits, sizes were slightly larger; however, three carat rough specimens were considered rare (Bauer, 1905, p. 359). They were also described as being ⅛ to ¼ in diameter (but rarely over ⅓ in) with a few measuring ½ in or larger (Kunz, 1890, p. 80). Sterrett, reporting on the Arizona pyropes, said that those stones which would cut one to two carat gems were plentiful; those which would yield one carat stones were abundant; and those which would cut three carat gems were scarce (Sterrett, 1908, p. 26).

The smallness of pyrope specimens in Bohemia created premiums very early for large stones. The premium scale given by de Boodt is a reflection of this rarity (de Boodt, 1609). In modern times, however, there may or may not be a price premium, depending on size. If the specimen is very large, perhaps competing with the world's largest, then there will be a premium because of size. However, if the stone is large, but not a possible record-breaker, perhaps fifteen carats, then there is a consideration of beauty. Since dark-toned stones increase their darkness with progressively larger sizes (light must travel further in the stone, thereby increasing its absorption), a much stronger light is necessary to bring out the colour. Consequently, unless the tone is quite light, modern cutters have learned to cut shallow pavilions in an effort to reduce this distance that light must travel. Indeed, some custom cutters prefer smaller sizes of rough material for the same reason.

Nevertheless, large stones are still acquired by collectors. But the choice of a large specimen represents a selection based upon rarity rather than the quality of the brilliance. In some cases, the rarity of size is a highly desirable choice.

Pyrope values and pricing

Pyrope garnets have fluctuated in value over the centuries. In de Boodt's time, 1609, garnets were equal to rubies in value. Yet, fifty years later, Nicols reported that the ruby was roughly equal to diamond in value, but that garnets were about equal to spinels, which were sold for about one-half the price of diamonds (Nicols, 1652, pp. 58–65).

It is interesting to note that in addition to mentioning a size premium for stones, Nicols also gave a pyrope quality standard for his day: '. . . and so proportionately according to the greater weight, great will be the increase of their value and worth, but with this proviso, that their colour for their glory, be always the perfect colour of the Rubine: for it is the pure excellence of its colour and tincture that determineth its price' (Nicols, 1652, p. 65). The brightness of the red hue plus the size were the fundamental standards of pyrope value. Among modern gem connoisseurs and collectors, the same standards are still in effect (see *Figure 3.1*).

In the late 19th century the feverish demand for factory-made Victorian jewellery stimulated a rising price structure for fine pyrope garnets. Although they were not as highly priced as rubies during that period, they were certainly more expensive than pyropes in today's gem marketplace. Bohemian stones were selling for as high as £25; but South African pyropes were more expensive, at £10 to £12.10s per carat for stones of moderate size (Bauer, 1905, p. 360). The Bohemian stones at that price were presumably rough specimens in very rare size categories; the South African pyropes were apparently cut stones.

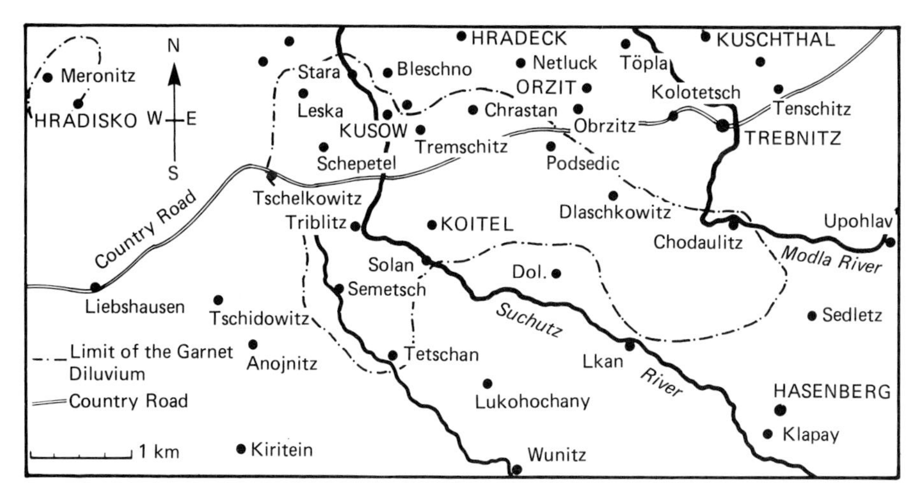

Figure 3.3 Map of the Bohemian garnet deposits (from Kunz, 1892)

Kunz also reported the price of the 'Arizona Rubies', as the pyropes from Arizona were called: the few finest pieces would fetch \$50 each; exceptional, but presumably smaller pieces would be \$5 each (probably uncut); while fine, one carat stones would sell for \$1 to \$3 each (Kunz, 1890, p. 81).

Certainly the high demand for fine gems and the rarity of exceptionally bright red colours from several mine sites created and maintained these high market values. It was only when the demand for gemstones, and/or their supply, was satiated or depleted, respectively, that these high values would diminish. If the prices reported here were to be compared to the value of the pound or dollar at this time, then they would be very high indeed (the \$1 stones might be equal to \$50 in today's inflated currency: a very high price for a 1 carat rough pyrope).

In the modern gem marketplace, rough pyropes have sold for as little as \$50 per kilo (2.2 lb). Even fine pyropes of two or three carats, cut might not be more than \$10 or \$20 per stone. Collector pyropes have been known to sell for as high as \$30 to \$50 per carat, but these would represent very rare exceptional gems.

Pyrope occurences

Earlier writers supposed pyrope to be found only in kimberlite, peridotite, or serpentine, a decomposed peridotite (Bauer, 1905; Kunz, 1892, etc). Certainly the early sources of Bohemia, South Africa and Arizona confirmed such opinions.

However, a study in 1938 suggested that pyropes could be found in other rock types as well. Having studied 223 garnets in the literature and calculating their end-member constituents, Wright (1938) reported pyropes were found also in various basic rock types, as well as eclogites, amphibole schists and biotite schists. Certainly the highest percentage of end-member pyrope components was found in peridotite and kimberlite (72.3%). Almandine also occurred in these rock types in varying percentages, in addition to grossular (although in much less amounts).

The association of pyrope with kimberlites has been useful in locating diamond sites both in Africa and in Russia. However, the presence of pyrope in kimberlite finds does not necessarily prove that diamond exists there, for no diamonds were found in the kimberlite dikes of Kentucky, although pyropes were discovered there (Kunz, 1904, p. 19).

Pyrope sources

Garnet is one of the most abundant gems on earth. Even a cursory glance at the gem garnets in the literature will reveal several dozen gem sources. However, it is sometimes most difficult to find eye-witness accounts, or even secondary authors who provide more than a brief comment on the site.

The modern researcher wants maps of the area, production figures, test samples, occurrences, methods of extraction, geological features, by-products, associated minerals, an adequate description of the gem sizes, colours and inclusions, as well as a complete history of the site: all documented with photos and bibliography.

However, most authors describing gem sites do not provide that level of data. Bibliographies are scant; eye-witness accounts are sometimes contradictory and secondary authorities are all too brief in their descriptions. In spite of these difficulties, however, the pyrope mines of Bohemia have had more press coverage than many other garnet locations. Bauer (about 1896) wrote with the authority of an eye-witness observer a very detailed account of the mines. Kunz wrote another detailed report somewhat earlier, in 1892, in which he reported on his trip to the area. Dr I.H. Oehmichen wrote an article in 1900 describing the geological features of the area. Most other writers quoted from these three; there are a few other general works, mostly in German, reporting technical aspects of the earlier periods. Sixteenth and seventeenth century accounts are very brief, simply noting the existence of the site, and not much more. One map was produced by Dr Wenzel Parek and appeared in Kunz' article in 1892 (*Figure 3.3*). It is adequate for describing the extent of the workings, but some key names of villages are missing, and there was no indication that the map was describing an area in northwest Czechoslovakia (see map of Bohemia, *Figure 3.4*).

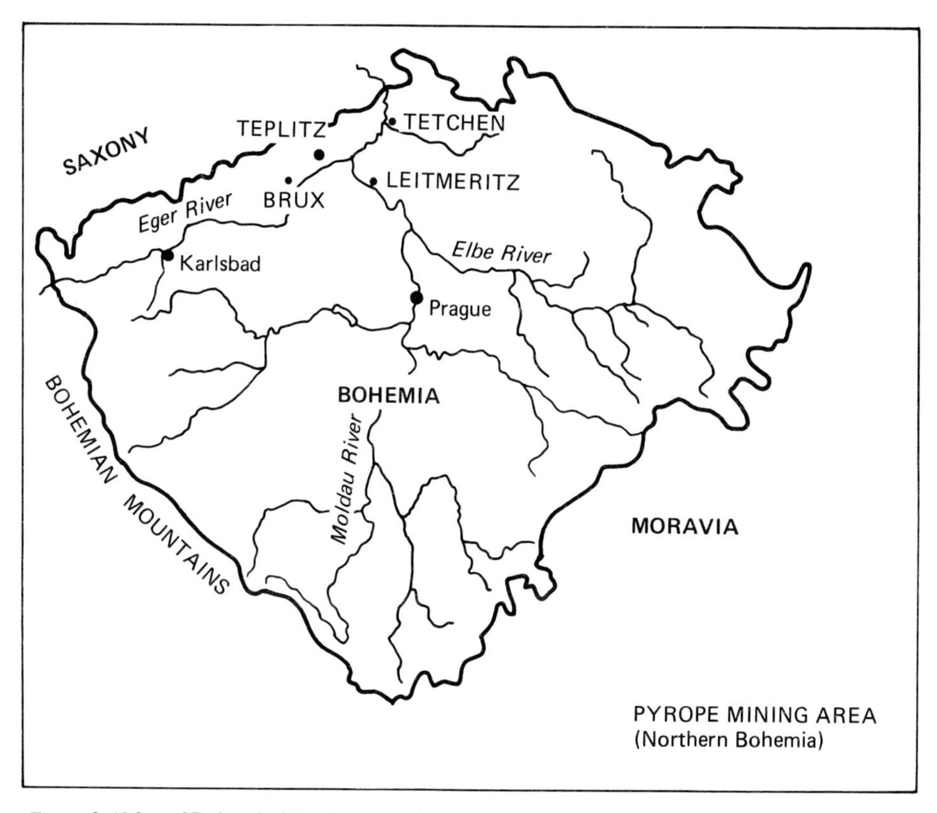

Figure 3.4 Map of Bohemia (Northwestern Czechoslovakia) from about 1900

The Bohemian mines

It is unknown when the first mining occurred in Bohemia but Bohemian garnet necklaces were found in grave sites dating back to the Bronze Age near the mining areas. Ball (1931) suggested that the mines were perhaps started about 1500 AD, but offered no documentation.

However, less than fifty years later, Agricola described the location of the mine site and said that it was 'five miles from Leitmeritz on the way to Trepnitz' (Agricola, *de Fossilium,* 1546, p. 172). Another Bohemian locality was mentioned near Brux, which is just over twenty miles due west from Leitmeritz. From later descriptions by Kunz (1892), this area was the exact location of the later mining activities; however, from Agricola's statement, the mining district was, of course, smaller in area.

In de Boodt's account of garnet, which he separated into Oriental and Occidental, Bohemian sites were listed under the latter category (de Boodt, 1609, p. 76). It was only a mining site, from all appearances, at this time.

Early in the next century (1715–1716), a mining company by the name of 'Granaten Gewerkschaft' received mining privileges in the area, and in 1723 a coin was struck from the metal in the garnet-bearing sand (Kunz, 1892, p. 6). Before the end of that century the mining area was transformed into a jewellery centre employing nearly a thousand people. This shift in emphasis was probably due, in part, to the first of the national exhibitions, held in 1791 in Prague (Kunz, 1892, p. 4). Besides gem cutting and goldsmithing, beads were also produced at that time; perhaps, but not certainly, they were faceted beads.

Bauer, in a very detailed description of the mines in Bohemia, suggested that 'after a period of decay', the centre was revived when mineral baths were started in Carlsbad, Teplitz and other places, resulting in a vast tourist trade which attracted people from all over Europe (Bauer, 1905, p. 357). Bohemian jewellery created in the area became very popular souvenirs and mementos. Bauer did not say when the baths were started, but possibly a date between 1723 and 1770 is most logical, especially with reference to the jewellery activity reported at the end of the century.

In the half-century between 1791 and 1850, there was an apparent steady growth in the area. However, from about 1850 until 1890, a boom occurred, and the industry flourished, undoubtedly stimulated by the popularity of Victorian jewellery. The new affluence was brought about by the Industrial Revolution, and, no doubt, to the vast supply of gemstones either found in, or brought to, Bohemia. Certainly the fact that the jewellery production industry was already in place in the area also stimulated and attracted new business. Both Bauer and Kunz reported the industry employed about 10 000 people in the late 1800s with the number of goldsmiths trebling since 1791.

It was also reported that many other gemstones were brought into the area from all over the world. Bohemia rapidly grew in dimension and developed into an international jewellery and cutting centre, somewhat reminiscent of the development of Idar-Oberstein in the mountains of western Germany. There was even a government school established in Turnau to teach gem cutting (Bauer, 1905, p. 357).

The industry was also diversified, for garnet jewels were created for watch bearings; garnet grains were used for counterpoises in delicate balances; abrasives were made from the garnets; and even garnet stones were used as ornamental gravel in garden walkways (Bauer, 1905, p. 357; Kunz, 1892, p. 4; and Farrington, 1903, p. 130).

The centre of the industry also expanded beyond the small area described by Agricola in 1546. A full seventeen towns and villages were listed as cutting centres, mining sites or jewellery production areas. Business activities were even conducted at Prague and its suburbs, some 40 miles to the southwest of the gem district (Kunz, 1892).

The jewellery created in Bohemia was of two types; cheap garnet jewellery for tourists and for export, and expensive gold fashion jewellery in the latest Victorian styles. Faceted garnet beads were also manufactured at this time. In the jewellery construction, Bauer mentioned the use of copper or silver foil backs and the 'mosaic style' jewellery, in which the metalwork existed for the sole purpose of displaying a pattern of gemstones and was blackened to minimize its appearance (Bauer, 1905, p. 359). An exquisite example of such jewellery is found in the Victoria and Albert Museum in London. This is a necklace of cabochon garnets in this style and is thought to be Bohemian, dating from about 1875 (Bury, 1982, Case 21, Board B, No. 1). The earrings do not match the necklace, but they are made in the same style.

Although Kunz described a decline in the district as early as 1903 (and some years earlier), the demise of the Bohemian industry was probably tied to many factors. Certainly the change in fashion occurring in the late 19th century was a factor. Perhaps the mines were also being depleted and competition with the 'more attractive' pyropes from the South African diamond fields also contributed. Nevertheless, the gemmological literary sources are quite silent thereafter. Even Webster's new fourth edition makes no mention of the fate of the Bohemian sites after the turn of the century.

It is quite probable that the area continued to be important as a mining centre until the depression hit in the late 1920s in Europe. Whether it recovered after that is questionable, but certainly with the area embroiled in World War II, if there was any mining done at all, it could only have been for industrial purposes. In the period subsequent to 1945 there was some revival in the area, particularly by residents who were mineral or specimen collectors. Several articles have been written in recent years, describing area features and its minerals (Rejl and Skalicky, 1977; Skalicky, 1978).

Other pyrope sources

In addition to the famous Bohemian deposits, there were many other sources for pyrope scattered throughout the world. In Europe, an early mention was made of deposits in the serpentines adjacent to Bohemia, in Saxony, within the forests of Zell and the village of Zoblitz (currently, East Germany; Mohs, 1825, p. 363). Other sites mentioned were Carinthia and Tyrol (modern Austria), Switzerland, and Hungary. But in 1825 the African pyropes were not yet known, nor the Arizona sites. However, the pyropes of Elie in Scotland were mentioned, as well as the garnets from Norway. The latter sites were said to be from Drammen and Arendale in Norway; current reports of pyropes are from the island of Otteroy, in western Norway (Hysingjord, 1971).

Even in 1867, Feuchtwanger was only just familiar with the US deposits on the eastern seaboard. He mentioned garnets in Buncombe County, North Carolina, Georgia, Massachusetts and New Hampshire, but did not distinguish whether they were pyrope or other garnets (Feuchtwanger, 1867, p. 249). In the Massachusetts site, cut gems were shown to Dr Feuchtwanger and they were described as 'beautiful' and 'precious'. Moreover, US Senator Clingman from North Carolina owned some fine gems, said to be derived from Buncombe County, North Carolina. The pyropes from Arizona and Africa were still unknown. Garnet sites in Moravia, Silesia (modern Czechoslovakia) and Siberia were added to the growing list of garnet sources, however (Feuchtwanger, 1867, p. 249).

The earliest mining activities in garnet fields of New Mexico, Arizona and Utah were accomplished by ants and scorpions. Large ant and scorpion hills were constructed by these busy creatures over garnet deposits. Garnet grains were pulled out of the sites by the insects, who viewed them as obstructions to their architectural planning. Indians came upon the hills and found many of these red pebbles to be of interest. Eventually, they traded them to the merchants, who sold them to travellers.

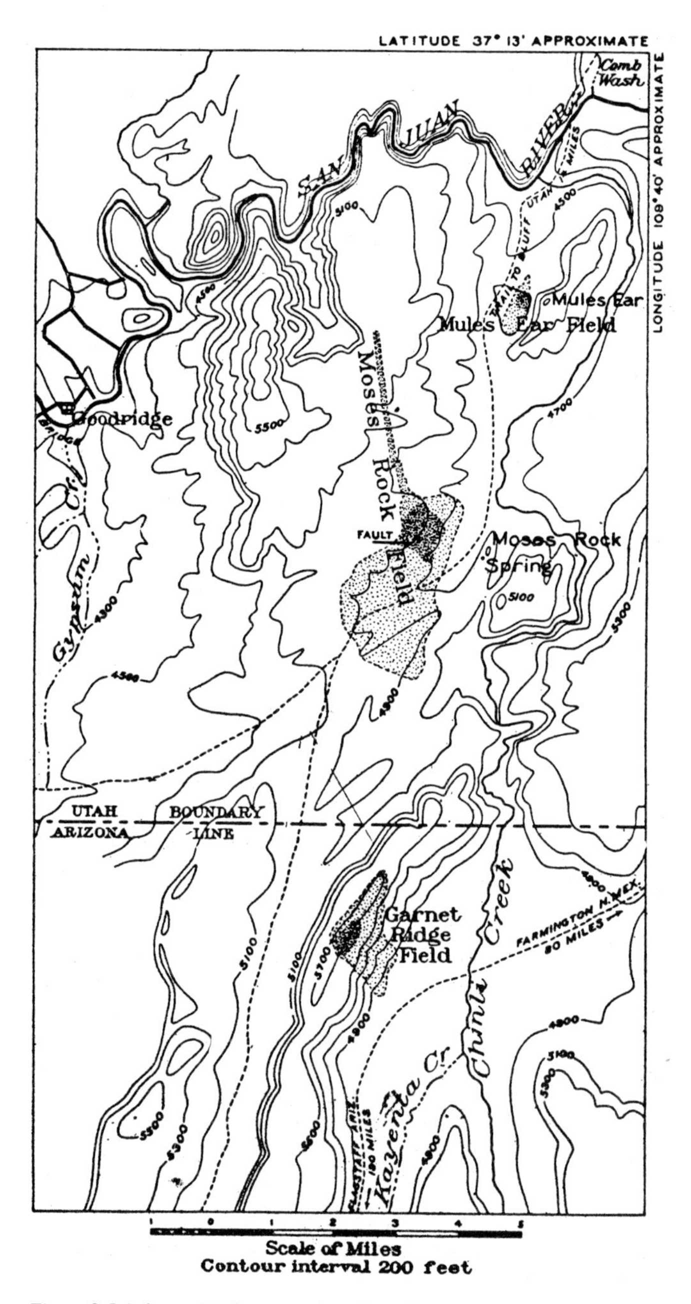

Figure 3.5 Arizona-Utah pyrope locations (from Gregory, 1916)

Considerable marketing of the gems was accomplished by traders and Indians; in fact, by 1908 Sterrett said that there was a scarcity of the material in the local markets and speculated that much of the material was being sold along the routes to Salt Lake City, Utah, and also to markets in New Mexico (Sterrett, 1908, p. 23). Sterrett's trip to the mines in northern Arizona proved very difficult, and when he finally reached the area, very little material was to be found, mostly in the shifting sand hills of the area (Sterrett, 1908).

The location of the northern Arizona deposits was precisely identified in 1916 to be on the border of Arizona and Utah (Gregory, 1916). The map located one deposit on the Arizona side of the border (*Figure 3.5*). It was called Garnet Ridge Field. Two other areas in Utah were called Moses Rock Field and Mules Ear Field. Apparently Sterrett's trip, reported in 1908, was slightly incorrect in pinpointing the exact location, for he corrected the report in a later writing to an area four miles from the original site, based largely on Gregory's earlier work in locating water in the area (Sterrett, 1911, p. 35). The colours of the pyrope gems were reportedly from a 'Burgandy wine' to a cinnamon hue (Sterrett, 1908).

Pyropes were also reported from New Mexico. They occurred largely on the Navajo Reservation, but they were also found in Sante Fe (Bauer, 1905). A large specimen of 11 ⅛ carats of fine gem quality was reportedly found in this area (Kunz, 1903, p. 42).

Many other sources of pyrope have been reported in the USA. Pyrope was found in Burke County, North Carolina, eight miles South East of Morganton along Laurel Creek; the colours were described as 'deep pink to rich red wine' (Sterrett, 1911, p. 26). Other areas in North Carolina yielded pyrope specimens, as well (Kunz, 1890, p. 81). Numbers of pyropes were found in Elliott County, Kentucky, which were highly regarded when cut into gems; the rough material from this site yielded stones from ⅒ in to ¼ in (Kunz, 1890, p. 81). The Stockdale kimberlite of Riley County, Kansas has also been reported (Bagrowski, 1941); many other sources could be mentioned if it were known whether or not the material was indeed pyrope, rather than almandine or some other garnet.

Another major source of pyrope was South Africa. It was associated with the diamond pipes and was mined as a by-product of the diamonds. Webster associated the fall of the Bohemian pyrope popularity to the fineness of the pyropes from South Africa (Webster, 1983, p. 172). The 'Cape ruby' was considered more valuable than any other garnet (Bauer, 1905, p. 360). Its colour was considered 'carmine-red' (orangish-red?), tinged with yellow, and moderate in tone (Bauer, 1905).

However, not all of the pyrope produced in the kimberlites were considered superior to the Bohemian pyropes. The diamond pipes in Russia also produce pyrope, perhaps a major deposit, for the pyropes are listed with other gems for commercial export (Novo-export, Catalogue). According to some authoritative reports, there is little pyrope in India, although pyrope-almandine and almandine garnets occur in profusion in many areas (Fermor, 1938, p. 55).

Pyrope garnet is also found in Australia. In the late 1800s a 'strike' of 'rubies' aroused such local interest that attempts to mine the material on a commercial scale were begun; unfortunately, after testing it, the ruby turned out to be garnet and the venture failed very suddenly (Males, 1976, p. 310). Pyrope also was early reported from Lowood in Queensland and near the diamond fields of Bingara in New South Wales (Males, op. cit.).

Summary of pyrope properties

Chemistry:	$Mg_3 Al_2 (Si O_4)_3$
Colouring agents:	Chromium, iron
Refractive index:	Pure:1.714; in nature: 1.730–1.750
Specific gravity:	Pure: 3.582; in nature: 3.65–3.80
Absorption spectra:	Iron-rich: 575 nm; 527 nm; 505 nm Chrome-rich: Strong diagnostic band between 520 and 620 mm; doublet of moderate strength at 687 and 685 and weak lines at 671 and 650 nm
Hardness:	7¼
Colours:	Slightly orangy-red, red, slightly violetish-red, in characteristic dark tones. High intensities common in the red hue
Dispersion:	0.022

Bibliography

AGRICOLA, G., *de Natura fossilium* (1546)

ANDERSON, B. W., 'Properties and classification of individual garnets', *The Journal of Gemmology,* VII, No. 1 (January, 1959)

ANDERSON, B. W. and PAYNE, C. J., 'The spectroscope and its applications to gemmology; part 13: absorption spectra of pyrope and topaz', *The Gemmologist* (September, 1954)

BAGROWSKI, B. P., 'Pyrope Garnet *vs* Ruby Spinel in Kansas', *American Mineralogist,* **26** (1941)

BALL, SIDNEY, H., 'Historical notes on gem mining', *Economic Geology,* **XXVI,** No. 7 (November, 1931)

BAUER, MAX, *Precious Stones,* translated by L. J. Spencer, Vermont and Tokyo (1970 reprint of 1905 edition)

BROOKINS, DOUGLAS G., 'Re-examination of pyrope from the stockdale kimberlite, Riley County, Kansas, *Mineralogical Magazine,* **36,** No. 279 (1967)

BURY, SHIRLEY, *Jewellery Gallery Summary Catalogue,* London (1982)

DANA, EDWARD SALISBURY, *Mineralogy,* sixth edition of James Dwight Dana, one volume, London (1896)

DE BOODT, BOETIUS, *Gemmarum et lapidum* (1609)

FARRINGTON, OLIVER CUMMINGS, *Gems and Minerals,* Chicago (1903)

FERMOR, L.DL., *Garnets and their Role in Nature,* Calcutta (1938)

FEUCHTWANGER, DR. L., *A Popular Treatise on Gems,* New York (1867)

FLOWER, MARGARET, *Victorian Jewellery,* London (1951)

GREGORY, HERBERT E., 'Garnet deposits on the Navajo Reservation, Arizona and Utah', *Economic Geology,* **11** (1916)

HANNEMAN, W. WILLIAM, 'A new classification for red to violet garnets', *Gems and Gemology* (Spring, 1983)

HYSINGJORD, JENS, 'A gem garnet from the island of Otteroy near Modle, Western Norway', *The Journal of Gemmology* (July, 1971)

JOBBINS, E. A., SAUL, J. M., STATHAM, PATRICIA M. and YOUNG, B. R., 'Studies of a gem garnet suite from the Umba River, Tanzania', *The Journal of Gemmology* (July, 1978)

KUNZ, GEORGE F., *Gems and Precious Stones of North America,* New York (1890)

KUNZ, GEORGE F., 'Bohemian Garnets', *Transactions of the American Institute of Mining Engineers* (February, 1892)

KUNZ, GEORGE F., 'Pyrope: Arizona and New Mexico', *Mineral Resources of the United States* (1903)

KUNZ, GEORGE F., 'Pyrope: Bohemia and Saxony', *Mineral Resources of the United States* (1903)

KUNZ, GEORGE F., 'Pyrope: Kentucky', *Mineral Resources of the United States* (1904)

MALES, P. A., 'Ruby corundum from the Harts Range, N.T.', *The Australian Gemmologist* (May, 1976)

MIERS, HENRY A., *Mineralogy* (1902)

MOHS, FREDERICK, *Treatise on Mineralogy,* London, 3 vol (1825)

NICOLS, THOMAS, *A Lapidary, or the history of pretious stones: with cautions for the undeceiving of all those that deal with pretious stones,* Cambridge (1652)

NIXON, PETER H., VON KNORRING, OLEG and ROOKE, JOAN M., 'Kimberlites and associated inclusions of Basutoland: a mineralogical and geochemical study', *The American Mineralogist,* **48** (September–October, 1963)

NOVOEXPORTS, *Catalogue of Semi-precious and Decorative Stones,* no date, received 1984, Novoexport 19, Bashilovskaya Str., Moscow A-287

REJL, DR. LUBOS and SKALICKY, JIRI, 'Gemstones of Czechoslovakia and their use in jewelry', *Lapidary Journal* (September, 1977)

SCHMETZER, KARL and BANK, PROF. DR. HERMANN, 'Garnets from Umba Valley, Tanzania. Is there a necessity for a new variety name?', *The Journal of Gemmology* (October, 1981)

SKALICKY, JIRI, 'Czech paradise and its gemstones', *Lapidary Journal* (September, 1978)

SKINNER, B. J., 'Physical properties of end-members of the garnet group', *American Mineralogist,* **41** (1956)

SMITH, HERBERT G. F., *Gemstones,* London (1977)

STERRETT, DOUGLAS B., 'Garnet', *Mineral Resources of the United States* (1908)

STERRETT, DOUGLAS B., 'Garnet:Utah', *Mineral Resources of the United States* (1908)

STERRETT, DOUGLAS, B., 'Garnet:Arizona', *Mineral Resources of the United States* (1911)

STERRETT, DOUGLAS B., 'Garnet:North Carolina', *Mineral Resources of the United States* (1911)

TRUMPER, L. C., 'Observations on Garnet', *The Journal of Gemmology* (October, 1962)

WEBSTER, ROBERT, *Gems,* fourth edition, London (1985)

WINCHELL, A. N. and WINCHELL, A., *Elements of Optical Mineralogy,* 4th Edition, New York (1933)

WRIGHT, W. I., 'The composition and occurrence of garnets', *American Mineralogist,* **23** (1938)

Chapter 4

Almandine

> From Asia's climes rich Alabanda sands
> The Alabandine and its name extends
> In fiery lustre with the Sard it vies
> And leaves in doubt the skilled beholder's eyes.
> Marbode, 11th century (King, 1867)

Almandine is undoubtedly the oldest garnet gem known to man. It certainly is the most abundant garnet in modern times; probably it was just as prolific in ancient times. Except for pyrope and perhaps grossular, the other garnets were too rare to be known, or their deposits were too remote for the times. Even pyrope deposits were probably unknown, particularly since Pliny failed to mention the Bohemian sites. Whether or not other garnet sources were known in the ancient world and unknown to the modern researcher is a possibility. However, even in that possibility, the occurrence was probably the ubiquitous almandine; certainly the ancient sites of India and Ceylon were largely almandine.

The term 'almandine', is derived from a small city in Asia Minor (modern Turkey), called 'Alabanda'. It was inland some miles from the port city of Miletus and was situated on a tributary of the Maeander, in easy reach of both Miletus and Ephesus. The town was small in size, but it enjoyed some stature in Roman times, if not before. Although there is some evidence that Alabanda was proverbial for its 'opulence and comfort' (Oxford Classical Dictionary, 1970), it also suffered much in history, including a sacking by Philip V of Macedon in his Carian expedition (201–197 BC), a period of independence under the Rhodian domination of the area, and very harsh treatment under a Roman provincial administration (Bean, 1976). The area was mentioned as a gem site by Theophrastus, and the city, a gem cutting centre by Pliny (see Chapter 1).

Although the site had been excavated in 1905, there was no direct evidence of a cutting trade in the city. However, only two small temples were excavated, and a large part of the site is still buried. An excavated theatre, another building, perhaps a bath structure, parts of an aqueduct, and a number of tombs just outside the city also remain. Interestingly the tombs, of the sarcophagus type, are said to be frequently inscribed with the professions of the deceased (Bean, 1976). The cutting centre, mentioned by Pliny, might be found in the ancient agora, or in private workshops established in other areas of the city.

Whether or not any remaining evidence of the cutting industry might be found even if the site were to be thoroughly excavated is a highly speculative, but a tantalizing prospect. Moreover, the area might be surveyed geologically in an effort to find evidence of any ancient garnet deposits, said by Pliny to be in nearby Orthosia (37:25).

Pliny considered the 'Alabandan' carbuncles to be a major source of the gemstone. It is significant that he described the finest quality of the carbuncle, the 'amethystizontas', just after mentioning Alabanda as a source (Pliny, 37:25). Pliny described the colour of amethystizontas as a violet-red. It is significant also that this colour is one of the major hues for modern almandine.

Although it was not defined as a violet-red gemstone by Marbode in the eleventh century, nor by Albert the Great in the mid-thirteenth century, Camillus Leonardus called the colour of 'Alabandina' bluish-red in 1520. However, he did admit that there was some confusion regarding the colour of the stone in his day. But the gem from Alabanda was not yet described as a garnet by Leonardus: it was simply a category of carbuncle.

A few years after Leonardus wrote his treatise, Agricola finished his important work on minerals (*de fossilium*). But he merely quoted Pliny that the 'alabandici' gems were rough ('scabricie') and dark. In the next century, Boetius de Boodt also quoted Pliny and earlier authors; but, he added that the Alabanda stones were classified between ruby and garnet, forming a separate category. He also claimed that the precious type came from India, and that they were cheaper than rubies (de Boot, 1609).

Later in the same century the Alabanda gemstones were reportedly scarce at any price and they were known only to a few people (Nicols, 1652). But the same author reported a new source: Pegu, a city (and kingdom) in southern Burma. They were reportedly found along the sands of the river (see *Figure 4.1*, which is a map dating from about 1870 showing this area). A very large gemstone was also mentioned from this location. It was owned by the king of Pegu (Nicols, 1652).

In the 19th century the Pegu source for almandine was still being reported (Mohs, 1825; Feuchtwanger, 1867). But in Bauer's work (1905) Dr. Noetling of the Indian Geological Survey argued that there were no garnets to be found in the whole of Burma. It was suggested that the fine violet-red almandines must have been brought from other sources, and Pegu, or its neighbouring city Syriam were simply trade outlets only. Consequently, the site was not well authenticated and garnets of this colour and species are not now known to exist in Burma, although some quantities of brownish-red almandines are found in both Burma and neighbouring Thailand. Just where the Pegu or 'Syrian' (a corruption from 'Syriam') garnets originated is still a mystery, but it is quite possible that they were brought overland from trade connections with India.

By the time of Mohs (1825) almandine was established as a separate species of garnet. It was described in very modern terms as an iron aluminum silicate with specific gravity measurements ranging between 4.098 and 4.208 (Mohs, 1825).

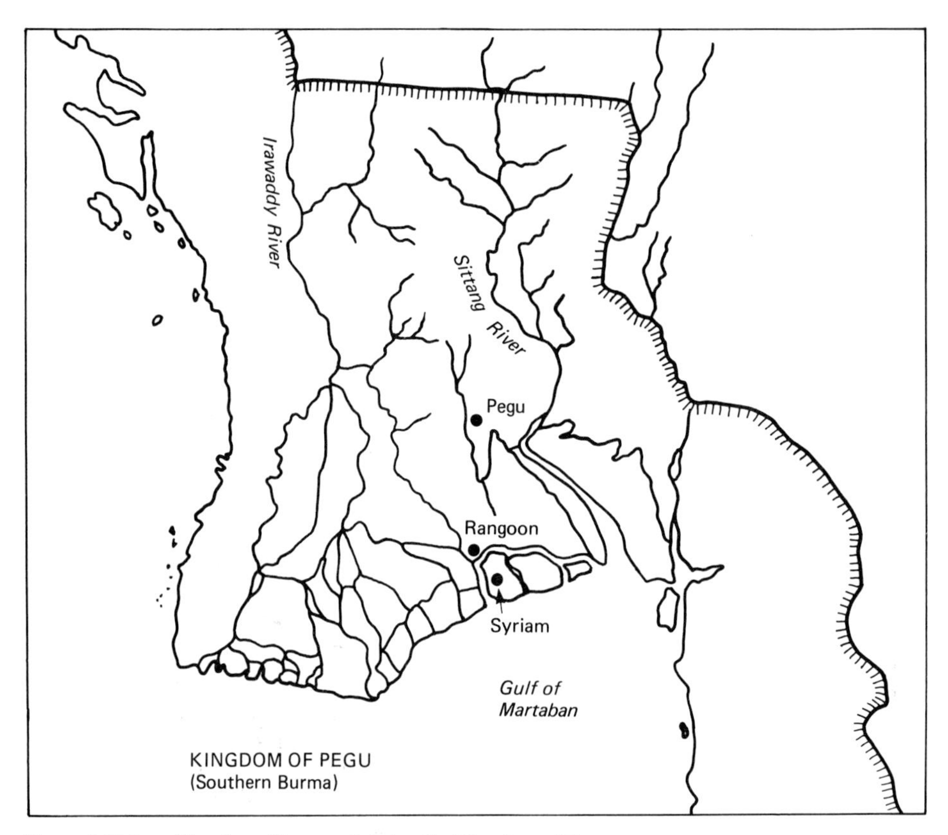

Figure 4.1 Map of Southern Burma, showing the Kingdom of Pegu

Almandine was still associated with the term 'precious garnet' and was reportedly found in the gem gravels of Ceylon as well as the river sands from Pegu. Later in the same century Feuchtwanger (1867) declared that the 'Syrian' garnet was blood-red in colour, while the 'vermeille' garnets were dark orange-red or brownish-red. The latter category was little valued because of the colour. 'Syrian' or 'Vermeille' became trade terms, associated with 'precious' and 'least valued' respectively. Bauer (1905) declared that the colour of the 'Syrian' garnet was violet-red, approaching 'that of the ruby or of the 'oriental amethyst'. Almandine, therefore, entered the twentieth century as 'precious' garnet from several sources and was categorized as garnet by chemical and specific gravity analysis.

Modern almandine chemistry: $Fe_3 Al_2 (Si O_4)_3$

Although Mohs reported chemical components of the various gemstones in his day, after the tradition of early mineralogists of the late 18th century, the testing methods gave incomplete data. Moreover, there were not enough samples tested from various sources. But by the end of the century chemical analysis was routine and it was beginning to contribute a much more thorough understanding both of the structure and classification of the garnets than earlier studies were capable of doing.

In modern almandine garnet, the chemical composition is $Fe_3Al_2(SiO_4)_3$. However, from an almandine chemical analysis exhibited in *Table 4.1,* a number of trace elements is also present. In addition, as in most minerals, there is a slight variation to be seen in the structural proportions. In this example the ferrous iron (Fe^{+2}) is accompanied by Mn, Mg, Ca, Ba, Na and K in the first position; traces of V^{+5} and Ti are found in the second position; and also some Al is combined with the Si in the third position.

TABLE 4.1. Almandine from eclogite; Garnet Ridge, Arizona showing chemical composition and structure. (From Watson and Morton, 1960, p. 273)

Refractive index: 1.797

Percentage of end-member molecules:

Molecule	%
Almandine:	69.5%
Spessartite:	2.4%
Pyrope:	12.0%
Grossular:	11.8%
Andradite:	4.3%

Chemical analysis (wet analysis):

Oxide	Value
SiO_2	37.53
TiO_2	1.38
Al_2O_3	20.65
Cr_2O_3	0.00
V_2O_5	0.02
Fe_2O_3	0.00
FeO	30.54
MnO	1.05
NiO	0.00
MgO	2.95
CaO	5.51
SrO	0.00
BaO	0.02
Na_2O	0.19
K_2O	0.02
P_2O_5	0.11
Co_2	0.00
S	0.00
H_2O+	0.03
H_2O-	0.00
Total:	*100.00*

Numbers of ions on the basis of 24 oxygens:

Ion	Number	Group
Si	5.982	Variation from the ideal 3:2:3 6.00 = 3
Al	0.018	
Al	3.861	4.04 = 2.02
Ti	0.166	
V^{+5}	0.002	
Fe^{+3}	0.000	
Fe^{+2}	4.069	5.92 = 2.96
Mn	0.142	
Ni	0.000	
Mg	0.700	
Ca	0.941	
Ba	0.001	
Na	0.060	
K	0.004	

From the end-member calculations, the dominant molecule for the Garnet Ridge example is definitely almandine (69.5%). But, a considerable amount of Mg indicates the presence of pyrope, which is usually present in all almandine. However, a high percentage of grossular is also present in this sample (11.8%), revealing another common occurrence. There are small amounts of spessartite (2.4%) and andradite (4.3%) in this sample as well. Since both pyrope and

TABLE 4.2. Almandine garnets reported in the literature, giving optical and calculated end-member constituents. The paucity of spessartite in the almandine here is not to be taken as an indication that little

	1	2	3	4	5	6	7	8	9	10	11	12	13
RI	1.775	1.775	1.778	1.779	1.783	1.783	1.787	1.787	1.790	1.790	1.792	1.794	1.795
SG	3.90	3.929	–	3.92	3.88	3.99	3.821	4.037	–	–	4.00	–	4.08
Major molecule end members (calculated)													
% Almandine	46	47.87	52.2	49	50	60	48	56.7	66.3	59.4	73	62.3	72
% Spessartite	–	2.30	1.6	–	–	–	–	7.7	1.9	1.8	–	2.3	10
% Pyrope	15	29.50	36.9	35	30	5	16	27.4	20.8	22.6	10	23.1	8
% Grossular	35	20.33	5.2	16	20	30	21	4.5	11.0	11.5	10	11.4	–
% Andradite	4	–	4.1	–	–	5	15	3.8	0.0	4.7	7	0.9	10
% Other	–	–	–	–	–	–	–	–	–	–	–	–	–

1. Gneiss, Burgess, Ontario, Wright (1938
2. Metamorphic rocks, Umba, Tanzania, Zwaan (1974)
3. Eclogite, Green Knobs, New Mexico, Watson and Morton (1969)
4. Schist, Markstay, Ontario, Wright (1938)
5. From anorthosite boulder, Adirondacks, N.Y., Wright (1938)
6. From limestone, Franklin, N. C. Zwaan (1974)
7. Glaucophane schists, Russian River, California, Pabst (1931); Recalculated and more Sp. noted, Pabst (1955)
8. Metamorphic rocks, Umba, Tanzania, Zwaan (1974)
9. Eclogite, Garnet Ridge, Arizona, Watson and Morton (1969)
10. Eclogite, Garnet Ridge, Arizona, Watson and Morton (1969)
11. Gneiss, Lake Harbor, Baffin Land, Wright (1938)
12. Eclogite, Garnet Ridge, Arizona, Watson and Morton (1969)
13. Pegmatitic granite, Jellicoe, Ontario, Wright (1938)

spessartite form a solid solution with almandine, it is not unusual to find them mixing in any proportions. But the andradite presence is an irregularity and the percentages, when present, generally are low.

An almandine property chart is presented in *Table 4.2*. It has been compiled from various sources which report RI and SG property values, in addition to end-member calculations derived from chemical analysis. From this data it can be observed that grossular occurs in most almandines in the chart in amounts as high as 30%, sometimes higher than pyrope. In such instances, the classification might be better called a grossular-almandine rather than a pyrope-almandine. However, only chemical analysis will reveal such a composition and since chemical testing is not a practical test for gemstones on a routine basis, gemmologists will probably continue to identify such a stone as a pyrope-almandine.

It can also be seen that the ferrous iron (FeO) (almandine molecular percentage) fluctuates from about 46% to about 80% in the almandines reported in *Table 4.2*. Pure almandine does not seem to exist in nature, yet higher percentages have been reported. In 1817 an almandine was said to contain 95.7% almandine; however, the calculations for Fe_2O_3 (ferric iron) were not determined (Deer, Howie and Zussman, 1962). Low figures are also found for the almandine end-member. The lowest value seen in *Table 4.2* is 46% (stone no. 1). A lower almandine has been reported at 38.54% FeO with no Fe_2O_3 (Deer, Howie and Zussman, 1962). The high or low ferrous iron content is also reflected in the refractive indices which fluctuate accordingly.

reported spessartite exists in almandine as a general rule. It is rather due to the absence of spessartite-rich pegmatites in this sampling

14	15	16	17	18	19	20	21	22	23	24	25	26	*Average* 1.7938
1.795	1.797	1.797	1.798	1.801	1.801	1.802	1.804	1.805	1.809	1.813	1.815	–	
4.02	–	4.12	4.00	4.12	4.04	4.09	4.09	3.97	4.10	4.22	4.20	4.235	
71	69.5	69.6	67	75.2	53.4	74	73.1	68	79	76.4	69.0	80.8	
4	2.4	3.5	3	4.2	1.8	5	3.3	2	6	5.6	27.0	2.0	
15	12.0	21.2	15	16.0	13.0	15	22.0	15	10	12.8	2.3	15.4	
10	11.8	5.7	5	3.2	23.8	6	1.5	5	5	3.1	1.7	–	
–	4.3	–	10	0.4	8.0	–	–	10	–	2.1	–	1.2	
–	–	–	–	1.0	–	–	0.1	–	–	–	–	0.7	

14. Gneiss, Sturgeon River, Saskatchawan, Wright (1938)
15. Eclogite, Garnet Ridge, Arizona, Watson and Morton (1969)
16. Crystalline schist, Fort Wrangel, Alaska, Pabst (1943)
17. Schist, Western Ontario, Wright (1938)
18. Silica-poor hornfels, Aberdeenshire, Rickwood (1968) recalculations of Deer, Howie and Zussman (1962)
19. Eclogite, Valley Ford, California, Rickwood (1968) recalculations of Deer, Howie and Zussman (1962)
20. Gneiss, New Hampshire, Wright (1935)
21. Banded garnet-biotite-sillimanite gneiss, Aberdeenshire, Rickwood (1968) recalculations of Deer, Howie and Zussman (1962)
22. Schist, Amisk Lake, Saskatchawan, Wright (1938)
23. Schist, Connecticut, Wright (1938)
24. Andestic tuff, Crinkle Crags, Cumberland, Rickwood (1968) recalculations of Deer, Howie and Zussman (1962)
25. Plutonic, Antarctic Peninsula, Vennum and Meyer, 1979; Center figures are reported here rather than the edge figures
26. Garnet-chlorite rock, Sweden, Rickwood (1968) recalculations of Deer, Howie and Zussman (1962)

The miscibility of almandine with other end-member garnets is also seen in *Figure 4.2*. While there is a continuous solid solution series between almandine and pyrope, and almandine and spessartite, yet there is some mixture into the ugrandite series also. There is commonly more grossular in almandine than either uvarovite or andradite, however.

Optical and physical properties

In gem almandine, the bottom limits for the refractive index and specific gravity are commonly reported as 1.78 and 3.95 (Webster, 1983). Any pyrope-almandine with properties above that limit are designated 'almandine'. Admittedly, these figures are arbitrary, since a chemical analysis is lacking. Yet, most of the stones in the chart (*Table 4.2*) are adequately classified using Webster's limits: only the first four are problematic.

The stones in *Table 4.2* have an average RI of 1.794 which is a common value for almandine. In addition, the end-member percentages which are high in andradite also seem to create higher refractive indices. Stone No. 19, for example, with high percentages of grossular and pyrope, seemingly should exhibit lower optical properties, but the presence of a considerable amount of andradite (8%) may contribute to the higher properties.

In the situation of separating the almandine from spessartite, neither the refractive index nor the density is useful, for there is considerable overlapping in

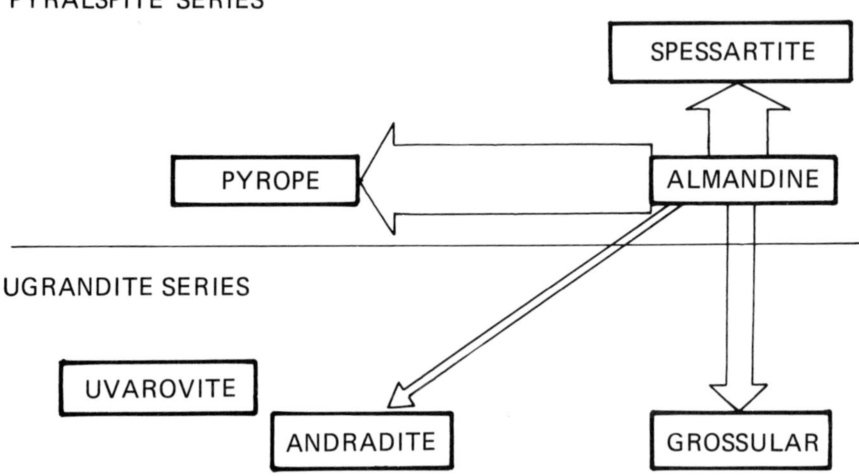

Figure 4.2 The miscibility of almandine is exhibited in relation to other end-member garnets. It can be seen that almandine mixes freely with pyrope and spessartite, but also with grossular and other ugrandite garnets

the two garnets. However, spectroscopic analysis and sometimes inclusion studies are effective in separating the two gems.

Almandines are the hardest of the garnets. They are commonly rated at 7½ on Mohs' scale. However, in a recent study of Ceylon almandines, a considerable hardness range was found in fifty-one cut gems. They were calculated to range from 6.81 to 7.48 in hardness on Mohs' scale (Arbunies, 1975). The range of hardness is also indicated by the reports of the special hardness attributed to the almandines from the Barton Mine in New York State (Hoffman, 1979).

Almandine absorption spectra

The characteristic iron bands in almandine have long been reported in gemmological literature, dating back to Church, 1866. The three main bands occur as a very strong band at 505 nm, as a moderate band at 527 nm, and as a strong band at 575 nm. These bands vary in strength with the percentage of the iron content in the sample. Lower percentages of iron diminish their visibility in the spectroscope somewhat; yet even in the low property pyrope-almandines, these three bands can still be seen. In the high-property almandines, they are quite strong, and other bands may also be seen.

The additional lines are found at 617 nm in the orange, and six lines in the blue to violet, at 476, 462, 438, 428 404 and 393 nm (Anderson and Payne, 1955). Since many of these occur in the dim region of the spectroscope, they are somewhat difficult for the novice to locate; but, they are easily seen with spectroscopic photography. Nevertheless, the three main bands are considered diagnostic for almandine, whether or not they are accompanied by the additional bands.

Almandine colours

Even a cursory glance at the literature reveals colour variations in almandine ranging from orange-red (or brownish-red) to red to violetish-red. The 'precious' almandine of the 19th century was considered violetish-red, a colour said to be in vogue at the time in both America and Europe (Bauer, 1905). The 'non-precious' vermeille was reported to be brownish-red or orange-red in hue. Colours described in modern times do not differ significantly from those above.

Almandines from Lapland are said to be 'greyish-rose' (colour 12B6, Methuen Book of Colour, 1978) and brownish-red (colour 9C8, Methuen), equivalent to violetish-red and orange-red, respectively (Hornytzkyj and Korhonen, 1980). The almandines from the Antarctic Peninsula are reportedly 'red-brown' in colour (Vennum and Meyer, 1979). Sri Lankan almandines are 'purple-red' (Arbunies *et al.*, 1975).

It should be stressed that the colours reported above do not necessarily apply to all garnets from the site: only those selected for study. Furthermore, the colours are determined in most cases from rough samples, or rough samples with polished windows, rather than faceted gemstones. Therefore, there could be some variation in colour due to the method of viewing. However, in our experience with almandine, the broad categories of orange-red to red to violet-red certainly are adequate in describing the general colours of the gems.

Pure almandine synthetics are 'cinnamon-brown' in colour (Deer, Howie and Zussman, 1962). If the iron is indeed the colour-producing element in almandine then one might expect a cinnamon-brown colour in natural almandines which approach the pure state. However, no such conclusion can be reached without studies of the chemical nature of almandine and its rôle in colour. There are many trace elements that could alter the colour in almandine garnets; furthermore, the influence of the trace elements is compounded with the contribution of both iron and manganese of the major end-member constituents. Nevertheless, almandine is classed as an idiochromatic gemstone, since it does not produce a colourless sample in its pure state, and since its major colour producer is the ferrous iron of its basic bulk chemistry.

The tones of the almandine tend to be dark. Cutters and jewellers have developed several methods to lighten its tone. One modern technique is to facet the almandine with a shallow pavilion, thereby decreasing the distance light must travel to pass through the gem. However, such a technique can also create another problem if the last rows of facets are too shallow for the optics. In such a case, light 'leaks' out of the bottom of the stone and a 'window' area appears. This area depends upon transmitted light reflecting off the object beneath the stone to produce colour. The 'window' area is often very different from the colour produced around the crown facets, which colour is produced by internal reflection rather than transmission.

Another method of lightening a dark almandine is to cut a cabochon with a curved, undercut bottom. This technique considerably lightens the tone, and has been used for centuries. Indeed, the term 'carbuncles' became synonymous with such stones in the last century.

Still another technique to lighten a dark almandine dates back to Pliny's day and involves the use of foil backing. A reflective metallic backing situated underneath the faceted gem or cabochon reflects much more light back into the viewer's eyes than those without backing. The appearance of the gem is definitely improved with such a process.

Intensities of the almandine hues tend to be moderate to dull. Yet, in some rare instances, bright to vivid hues are encountered.

Almandine inclusions

Almandine exhibits very many inclusion-types (see *Table 4.3*). Many of these are characteristic of the gemstone; but perhaps few are characteristic of the source. For purposes of inclusion studies, it seemed appropriate to consider the almandine and the intermediate pyrope-almandine as a single group, for there does not seem to be any demarcation between them regarding the nature of the inclusions.

Among the most common inclusions in almandine are the ubiquitous 'needles', which are rutile crystals. They are often oriented so that they intersect each other on the same plane in angles of 70° and 110°. When they occur in profusion in a faceted stone, they produce a darker tone. The light travelling through the gem is diffused; therefore the overall colour appearance is weakened. If the needles occur *en masse*, then the rough stone is usually cut into a cabochon which reveals an attractive four-ray-star. The rutile acicular inclusions in faceted stones may be long or short; they may intersect at the angles above, or be positioned in a seemingly random fashion. Their presence in faceted almandine is not detrimental to the gem's appearance unless they occur in bundles. They also vary from slightly coarse to fine in thickness.

In addition to the rutile needles, almandine often serves as the host for other rutile crystals (Zwaan, 1974). The rutiles are tetragonal and exhibit high refractive indices (2.62-2.90), thereby standing out in high relief in the host garnet. Their lustre is also high (metallic-adamantine).

A very common inclusion in almandine is apatite. This mineral crystallizes in the hexagonal system and is sometimes seen as fairly well-developed crystals in the host almandine. Its optics are lower than the rutile, and also lower than the host garnet, exhibiting an RI of only 1.630–1.648. Also, the lustre is only just vitreous.

Other common crystal inclusions in almandine are zircon, spinel and quartz. Zircon, due to its disruptive force as a metamict mineral, often is accompanied by telltale stress fractures in a halo-like form around the zircon crystal. Biotite and phlogopite flakes from the mica group are also reported to be common (Gübelin, 1979). These mica inclusions often occur as flat platelets and in groups.

Two-phase inclusions, dendritic inclusions and fibrous inclusions (similar to the byssolite fibres in demantoid) have also been reported in almandine, although they are rare. Other crystal inclusions that have been identified by X-ray powder photography (Dunn, 1975; Zwaan, 1974) are pyrrhotite (magnetic pyrites), ilmenite (titanic iron ore), fluorapatite, chalcopyrite, plagioclase and muscovite

TABLE 4.3. Inclusions in almandine and intermediate pyrope-almandine

Source and date	*Location*	*Inclusions found*	*Gem I.D.*	*Method of inclusion I.D.*
Martin (1970) and others	N.Y.	Two-phase inclusions	Rhodolite	Appearance
Dunn (1975) and others	N.Y.	Apatite	Almandine	X-ray
Zwaan (1974) and others	Umba, Tanzania	Rutile needles	Almandine	Appearance
Zwaan (1974) and others	Umba, Tanzania	Rutile crystals	Rhodolite	X-ray
Zwaan (1974) and others	Umba, Tanzania	Feathers, liquid	Almandine	Appearance
Zwaan (1974) and others	Umba, Tanzania	Zircon haloes	Rhodolite	Appearance
Zwaan (1974) and others	Umba, Tanzania	Platy inclusions (mica group)	Rhodolite (1.753)	Appearance
Zwaan (1974) and others	Umba, Tanzania	'Smooth tubes with rounded edges'	Rhodolite (1.762)	Appearance
Dunn (1975) and others	N.Y.	Pyrrhotite	Almandine (1.764)	X-ray
Dunn (1975) and others	Idaho	Ilmenite	Almandine (1.808)	X-ray
Dunn (1975) and others	Idaho	Fluorapatite	Almandine (1.808)	X-ray
Dunn (1975) and others	N.Y.	Chalcopyrite	Almandine (1.764)	X-ray
Dunn (1975) and others	N.Y.	Plagioclase	Almandine (1.764)	X-ray
Dunn (1975) and others	N.Y.	Acicular Cavities	Almandine	Appearance
Rouse (1984)	Madagascar	Fingerprint	Pyrope-almandine (1.759)	Appearance
Hornytzkyj and Korhonen	Finland	Monazite	Almandine	Morphology
Hornytzkyj and Korhonen	Finland	Actinolite	Almandine	Morphology
Zwaan (1967)	Sri Lanka	Muscovite	Almandine (4.176 S. G.)	X-ray
Zwaan (1967)	Sri Lanka	Sphalerite	Almandine	Not mentioned
Martin (1970)	Unknown	'Comet-like needles	Rhodolite	Appearance
Patrick (1972)	Unknown	Dendritic inclusion	Almandine	Appearance
Journal of Gemmology (1949)	Unknown	Fibrous inclusions	Almandine	Appearance
Gübelin (1979)	Unknown	Biotite and phlogopite flakes	Almandine	Appearance
Gübelin (1979)	Unknown	Quartz	Almandine	Appearance
Gübelin (1979)	Unknown	Spinel	Almandine	Appearance

(common mica). Sphalerite, a high-property zinc mineral is also mentioned (Zwaan, 1974).

In addition to crystals, almandine often hosts liquid-filled inclusions. They have been described as 'liquid feathers' or 'fingerprints' when they occur either as an ill-defined veil or as a tightly composed fingerprint pattern, respectively. Although the fingerprint inclusions are rare, they are reported from several sources (Ceylon, Trumper, 1952; Madagascar, Rouse, 1984).

Almandine sizes

Unlike pyrope, which is usually very small in size, almandine is often large. One of the most famous large stones is one in the Kunz collection. It is a well-formed trapezohedron weighing 4.4 kg, found in New York City at Thirty-fifth and Broadway (Manchester and Stanton, 1917; Kunz, 1890). From Ruby Mountain in Colorado specimens as large as 6.3 and 6.6 kg were reported (Kunz, 1890). Large crystals are not uncommonly reported among the numerous worldwide sources of almandine.

In the consideration of gem quality specimens, however, smaller sizes are more commonly encountered and preferred. Indeed, large cut stones increase the distance light must travel, thereby increasing the depth of the tone. Consequently, smaller gems are more suitable for displaying the colour and for mounting in jewellery; and large gem almandines carry no special value premium for these reasons.

Some large specimens of almandine have been made into drinking vessels. Pliny reported vessels of 'lychnis' (garnet) and another stone from Egypt, near Thebes (also garnet) which could hold a pint of liquid. However, they were full of veins, brittle, and perhaps quite dark in tone (Pliny 37:30), conditions quite common in large almandines.

Other large specimens are also found in the literature. King (1866) reported: 'I have myself seen a small antique bowl (Graeco-Roman) of the size of a Chinese teacup formed out of a single garnet, and bearing its owner's name κόδρος (Kodros), engraved on the inside'. He reported another large specimen, a slab 2½ by 1½ in, carved with a long Gnostic formula on both sides; this work was dated from the third century AD and was assigned to the Alexandrian school (Kunz, 1867). Goodchild (1908) also mentioned large specimens: 'Bardot, in his inventory of the French jewels, made in 1791, mentions two cups made of garnet of three inches in height, and several small ones'.

From the examination of many almandine deposits in the literature, there seems to be a relationship between size and source. Some mines produce only small stones; others produce large sizes as a common occurrence. For example, the famous Barton Mine in New York state has produced almandines whose diameters range from 2 to 6 in in considerable volume; occasionally garnets can occur there in diameters from 1 to 2 ft (Miller, 1938). There may also be a relationship between habit and size, as demonstrated by Pabst (1943).

Almandine occurrence

In contrast with pyrope, almandine commonly occurs in many rock types. The most common occurrences are mica schists and gneisses. However, they also occur in pegmatites, granite, biotite schists, amphibole schists, eclogites, various basic rocks and even kimberlites. In Wright's study (1938), the almandine end-member percentage was the highest proportion (73.0%) among biotite schists than any other rock-type. The smallest percentage of almandine was in kimberlite at 13.4% (Wright, 1938).

Almandine sources

Having such a wide distribution of occurrences, it is not surprising to find numerous sources of almandine around the world. The historical sources of Alabanda, Syriam (Pegu) and India have already been mentioned. Although the Alabanda source may not have been a mining area, only a cutting and trading centre, it is possible that almandines may have occurred in the district, only to be mined out in ancient times. At any rate, gem almandines were acquired from India during Greek and Roman times, if not before.

The major historical Indian sources were principally in the north, in Rajputana. However, almandines were also found in the gem gravels of Ceylon and many other locations in India (see Iyer, 1961, and the *Wealth of India*, 1956 for a detailed account). The large area of western India seems devoid of garnet occurrences, however (Fermor, 1938). The Indian mines have been prolific producers of almandine; the area was the major source for Roman garnets as well as the modern world, particularly since 1500. Like many other large deposits, garnets from India have also been utilized on a large scale as abrasives, as well as gemstones and other ornamental objects.

Of the many sources of almandine in the USA, several are well known. The Barton Mine in the Adirondack mountains of New York State is quite famous; it was started by Henry Barton in the late 1880s. Mr Barton learned about the deposit when a customer showed samples of the material to a jeweller where Mr Barton was apprenticed. Years later when he opened up a woodworking supply and abrasives shop in Philadelphia, he remembered the garnets and sought out the source.

The garnets proved to be far superior abrasives than the crushed glass sandpaper he was then selling. The deposit proved to be prolific in quantity and quality and was in continuous operation until 1984. The deposits additionally provided the lapidary and faceter with numerous samples of gem material over the years, although the colours are generally dark in tone. The site is well-known for producing garnets of superior hardness and considerable size (Hoffman, 1979).

In addition to the deposits in New York State, almandines have also been found in New York City. The sites have been described as occurring along Broadway, at 35th St., 65th St. and 165th St. The garnets were mostly almandine although spessartite and grossular have also been found there (Manchester and Stanton,

1917). It was at 35th and Broadway where the famous 4.4 kg near-perfect trapezohedral almandine was discovered and later acquired by Kunz. Stanton is reported to have found 'hundreds of fine dodecahedrons with trapezohedral truncations, mostly from three-eighths to one inch in diameter' (Manchester and Stanton, 1917). Another large almandine weighing 4.8 kg was also found at 165th Street, but the specimen was not well-formed.

Other almandine deposits in the USA, producing a substantial quantity of abrasive garnet material, are the North Carolina sites, located in Burke, Caldwell and Catawba Counties. These garnets were unusual in that they were coated with a brown covering of limonite (due to a superficial alteration). They varied in colour from violet-red to purplish-red and occurred in some large sizes. Some crystals were reportedly found in sizes up to 9 kg each, and, although not of gem quality, might have been useful for dishes or cups measuring from 3 to 6 in across. They also displayed a peculiar play of colour that was due to inclusions (Kunz, 1890).

Pennsylvania also produced some almandine garnets. The discovery was situated in eastern Pennsylvania, where almandines of a fine violet colour were found (Sterrett, 1911).

Southern Colorado has also been mentioned as a source of almandines. Kunz (1890) reported a remarkable deposit of almandine at Ruby Mountain in Chaffee County. Fine crystals occurred in a bed of green chlorite and varied in size from 28 grams to several kilograms each, two specimens weighing 6.4 kg each. The deposit was chiefly utilized for mineral specimens. Kunz said that a minimum of five tons of material had been used for this purpose. Unfortunately they were not of gem quality, since they were not transparent. Another site in Colorado was reported to be very near to Canon City. This deposit did produce gem garnets which were described as a beautiful red to pinkish-red; gems were cut to about two carats in size. Two mining claims were taken out for this material which occurred along Grape Creek (Sterrett, 1908).

Almandine garnets were reported early from Idaho. Kunz (1890) mentions gem quality almandine from the gravel of placer mines in Lewiston, Idaho. The rolled and pitted grains were from 1⁄16 in to 1 in diameter and they were described as useful for watch jewels or gemstones. Along Emerald Creek in Latah County, another major deposit of garnet was discovered. The garnets were utilized primarily as abrasives, and production was high from World War II to the present time. The production reached 10 000 tons per year in 1963, and it was estimated that there was enough material in the area to last into the 1990s (Agee, 1965). Stones from the deposit were largely opaque, although occasional faceting grade gems could be found. However, in gemmological circles, the deposit is chiefly known for its production of star almandines.

Many other sources of almandine have been discovered in the USA. However, most sites are either small, of little gem significance, or historical deposits no longer worked. The almandine sites from the Columbia River (Goodchild, 1908), from the mouth of the Russian River in California (Pabst, 1931), from Ely, Nevada (Pabst, 1938), from eclogite inclusions at Garnet Ridge, Arizona (Watson and Morton, 1969), and from many other sites are examples of such sources.

However, an important source of almandine was also found in Alaska. Kunz

(1890) reported that the crystals from the Fort Wrangel, Alaska area were quite perfectly formed, and were found in great quantities near the mouth of the Stikeen River. As crystal specimens they were greatly admired. Indeed, the thousands of samples said to have been brought out by ship bear witness of an important commercial activity. A number of such specimens was reportedly taken to the United States National Museum.

There are minor sources of almandine in both Canada (Baffin Island, etc) and Mexico (Baja California). The Swat Valley in NW Pakistan is another source of almandine, in addition to the orange-red almandines of Burma and Thailand. Other sources include Zambia, along the Mazabika River, Namaputa in the Lindi Province of Tanzania, Austria, Bohemia and many other locations. The Bohemian deposits were relatively minor and did not produce gem material in any significant quantities.

Brazil also produces gem almandine. It occurs in the gem-rich state of Minas Gerais, as well as in Bahia and the state of Rio de Janeiro. It also occurs in the remote areas of Antarctica, in Finland, in Spain and Australia.

In Australia, there is a significant source in the rivers of the Northern Territory of South Australia. At one time, in the late 19th century, the deposit triggered a 'ruby' rush, as they were thought to be rubies. Before they were officially identified as almandine, not less than twenty-four ruby companies were formed, only to collapse with the news that they were garnet rather than ruby (Bauer, 1905).

Greenland also is a reported source for almandine. The region east of Godthaab is often mentioned, as well as the west coast from Disko Bay South, and many other areas (Sinkankas, 1959). Garnet masses up to 25 kg have been found in the Godthaab district.

Almandine phenomena stones

There are several interesting phenomena associated with almandine. One is the asterism due to the silk, or rutile acicular crystals that can saturate the gemstone. They are found in India and Idaho, perhaps also Sri Lankan. Generally they exhibit a four-ray star. But six-ray stars and stones with four and six rays do occur. Very interesting four-ray stars exist with a six-ray star at the extremity of each leg of the four-ray star. Some spheres of this material which show the full effect of this fascinating phenomenon have also been cut (Walcott, 1937).

Another unusual phenomenon in almandine has been the report of a 'play of colour' (Kunz, 1890). In this case, the cause for this phenomenon was suggested to be due to the inclusions in the stones. Unfortunately, no later writer was found who mentioned it, so it is probably an isolated instance from this source only (North Carolina).

Summary of almandine properties

Chemistry: $Fe_3 Al_2 (Si O_4)_3$
Colouring agents: Iron
Refractive index: Pure:1.830; in nature: above 1.780
Specific gravity: Pure: 4.318; in nature: above 3.95
Absorption spectra: Main diagnostic lines:
505 nm Very strong band
527 nm Moderate band
575 nm Strong band
Other lines sometimes seen:
617 nm, 476 nm, 462 nm, 438 nm, 428 nm, 404 nm, 393 nm
Hardness: 7½
Colours: Orangy-red, red, violet-red, in characteristic dark tones; high intensities are sometimes found in the red and violet-red hues
Dispersion: 0.027

Bibliography

AGEE, LEON, M., 'Asterism in Garnets', *Lapidary Journal,* **19** (November, 1965)

ANDERSON, B. W. and PAYNE, C. J., 'The spectroscope and its application to gemology. Part 19: absorption spectrum of almandine garnet', *The Gemmologist* (March, 1955)

AGRICOLA, G., *de Natura fossilium* (1546)

ARBUNIES-ANDREU, BOSCH-FIGUEROA, J. M., FONT-ALTABA, M. and TRAVERIA-CROS, A., 'Physical and optical properties of garnets of gem quality', *Fortschr. Miner,* **52** (1975)

BAUER, MAX, *Precious Stones,* translated by L. J. Spencer, Vermont and Tokyo (1970 reprint of the 1905 edition)

BEAN, G. E., 'Alabanda', *The Princeton Encyclopaedia of Classical Sites,* edited by Richard Stillwell, Princeton, New Jersey, Princeton University Press (1976)

DE BOODT, BOETIUS, *Gemmarum et Lapidum* (1609)

DEER, N. A., HOWIE, R. A. and ZUSSMAN, J., *Rock-Forming Minerals. Ortho and ring silicates,* **1,** London (1962)

FERMOR, L. L., *Garnets and their Role in Nature,* Calcutta (1938)

FEUCHTWANGER, L., *A popular Treatise on Gems,* New York (1867)

FOLINSBEE, ROBERT E., 'The chemical composition of garnet associated with cordierite', *The American Mineralogist* (26 January, 1941)

GOODCHILD, W., *Precious Stones,* New York (1908)

GUBELIN, E., *Internal World of Gemstones,* Zurich (1979)

HOFFMAN, ALAN C., 'The garnet-empire state's gem of many facets', *Lapidary Journal* (August, 1979)

HORNYTZKYJ, S. and KORHONEN, K. T., 'Notes on the properties and inclusions of garnets from Lapland, Finland', *The Journal of Gemmology,* **XVII,** No. 3 (July, 1980)

IYER, L. A. N., 'Indian Precious Stones', *Geological Survey of India,* Bulletin No. 18 (1961)

KING, C. W., *Antique Gems: their Origin, Uses and Value,* London (1865)

KING, C. W., *The Natural History of Precious Stones,* London (1867)

KORNORUP, A. and WANSCHER, J. H., *Methuen Handbook of Colour,* Norfolk (England, 1978)

KUNZ, GEORGE FREDERICK, *Gems and Precious Stones of North America,* New York (1890)

LEONARDUS, CAMILLUS, *The Mirror of Stones,* English translation 1750 from the original Latin in 1520

MANCHESTER, JAMES G. and STANTON, GILMAN S., 'A discovery of gem garnet in New York City', *The American Mineralogist,* **II,** no. 7 (July, 1917)

MILLER, WILLIAM J., 'The garnet deposits of Warren County, New York', *Economic Geology,* **7** (1912)

MILLER, WILLIAM J., 'Genesis of certain Adirondack garnet deposits', *The American Mineralogist,* No. 6 (June, 1938)

MOHS, FREDERICK, *Treatise on Mineralogy,* 3 vol, London (1825)

NICOLS, THOMAS, *A Lapidary, or the history of pretious stones: with cautions for the undeceiving of all those that deal with pretious stones,* Cambridge (1652)

Oxford Classical Dictionary, 2nd edition, Oxford, 1970, 'Alabanda'

PABST, ADOLPH, 'Garnets from vesicles in rhyolite near Ely, Nevada', *The American Mineralogist,* **23,** No. 2 (1938)

PABST, ADOLPH, 'Large and small garnets from Fort Wrangell, Alaska', *The American Mineralogist,* **28,** 4 (1943)

PABST, ADOLPH, 'Manganese content of garnets from the Franciscan schists', *The American Mineralogist,* **40,** 9–10 (1955)

PABST, ADOLPH, 'The garnets in the glaucophane schists of California', *The Amer. Miner.,* **16,** 8 (1931)

PATRICK, S. G., 'Garnet inclusion', *The Australian Gemologist* (November, 1972)

PLINY THE SECOND, *Natural History,* Loeb Classical Library edition, translated by D. E. Eichholz, Harvard, 1962, Volume X, Books 36, 37

SHAUB, B. M., 'Paragenesis of the garnet and associated minerals of the Barton Mine near North Creek, New York', *The American Mineralogist,* **34** (1949)

SINKANKAS, JOHN, *Gemstones of North America,* Vol. I (1959), Vol. II (1976), New York

STERRETT, D. B., 'Garnet', *Mineral Resources* (1908)

STERRETT, D. B., 'Garnet', *Mineral Resources* (1911)

TRUMPER, L. C., 'Observations on garnet', *The Journal of Gemmology* (October, 1962)

'Unusual inclusions in almandine garnet', *The Journal of Gemmology* (January, 1949)

VENNUM, WALTER R. and MEYER, CHARLES E., 'Plutonic garnets from the Werner batholith, Lassiter Coast, Antarctic Peninsula', *American Mineralogist,* **64** (1979)

WALCOTT, ALBERT J., 'Asterism in garnet, spinel, quartz and sapphire', Geological Series of the Field Museum of Natural History, Vol. VII, No. 3 (December 28, 1937)

WATSON, K. D. and MORTON, D. M., 'Eclogite inclusions in kimberlite pipes at Garnet Ridge, Northeastern Arizona', *The American Mineralogist,* **54** (January–February, 1969)

The Wealth of India, 'Garnet', Vol IV, New Delhi (1956)

WEBSTER, ROBERT, *Gems,* fourth edition, London (1983)

WESTROPP, HODDER M., *A Manual of Precious Stones and Antique Gems,* London (1874)

WRIGHT, W. I., 'The composition and occurrence of garnets', *The American Mineralogist,* **23** (1938)

ZWAAN, P. C., 'Garnet, corundum and other gem minerals from Umba, Tanzania', *Scripta Geol.,* **20** (1974)

ZWAAN, P. C., 'Solid inclusions in corundum and almandine garnet from Ceylon, identified by x-ray powder photographs', *The Journal of Gemmology,* **10,** 7 (1967)

ZWAAN, P. C., 'Some notes on the identification of the pyrope-almandine garnets', Koninks. Nederl. Akademie van Wetenschappen-Amsterdam. Reprinted from Proceedings Seriel B. 64, No. 2 (1961)

Chapter 5

Intermediate categories

Since there are a number of garnets that are in solid solution with each other, there are several possibilities for garnet intermediates. For example, spessartine-almandine end-members are in solid solution, and they could and do, produce gemstones with properties midway between the two species. However, there are important differences between the intermediates of spessartine-almandine and pyrope-almandine.

First, there is a broad gap separating the RI (and SG) of the pyrope-almandine species, allowing for easy separation with gemmological instruments. The spessartine-almandine properties overlap each other, resulting in difficult separation tasks for the gemmologist.

Also, the pyrope-almandine intermediates seem to produce colours that, in general, stand out prominently against their end-members. This difference was noted by gem observers of both the 19th and 20th centuries. However, the colour distinction of the spessartine-almandine intermediates is not well known and generally blends gradually into the colours of the end-members. The rarity of other intermediate categories also discourages category definition.

Moreover, within this 'gap' between pyrope and almandine, stones were discovered in the late 19th century that seemed to the early gemmologists uniquely different in appearance from their end-members. The 'new' stones were remarkable both in their hue position and the tone of the hue. Light violet colours (lavender) and rich violet-red hues of medium tonal dimension were discovered in North Carolina. These stones stood out in high relief against the very dark gems of either pyrope or almandine. The excitement they aroused was profound in a day when such garnets were virtually unknown. Other garnets were also found in the same intermediate classification, but of a different colour; they also are part of this unique category of gems.

Historical

> 'The analysis proves that this garnet is not almandine nor wholly pyrope and is distinctive enough to merit a varietal name. We, therefore, propose the name of Rhodolite, from the two Greek words ῥόδον, a rose, λίθος, a stone . . .' Hidden and Pratt, on the introduction of rhodolite in 1889.

From this statement by Hidden and Pratt, the term 'rhodolite' came into the gemmological vocabulary. The 'new' gemstones did seem unusual when compared with the pyropes and almandines in vogue during the late 1800s. Rhodolite tones were not extremely dark; and the delicate rose-like colour was unique and very appealing. Furthermore, chemical analysis indicated that the gems were situated between the end-member species of pyrope and almandine, possessing one molecule of almandine and two molecules of pyrope. There is little doubt that the gemstones 'excited unusual interest and admiration' among the mineralogists and gem enthusiasts of the day.

The gems were found in Macon County, North Carolina, 'upon a brook known as Mason's Branch' (Hidden and Pratt, 1898). This stream flows into the Little Tennessee River (see *Figure 5.1*) and was perhaps the first deposit of rhodolite to

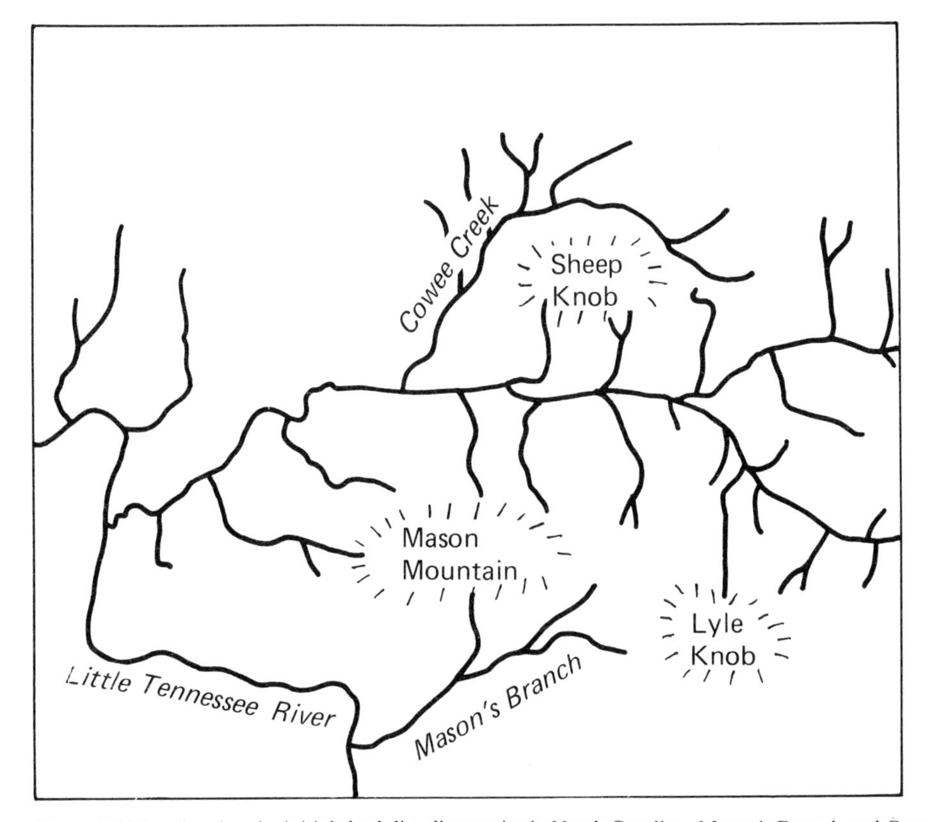

Figure 5.1 Map showing the initial rhodolite discoveries in North Carolina. Mason's Branch and Cowee Creek both produced the rhodolites, but only Cowee Creek produced corundum

be mined. However, there is some doubt on this issue, for there was another source of rhodolite just north of Mason's Branch in the Cowee Creek area where ruby was found with the rhodolite (Sinkankas, 1959). Exploration along Cowee Creek, according to Sinkankas (1959), led to the discovery of the rich Mason's Branch deposits which are reported by Hidden and Pratt, above. The Mason's Branch site was, however, known from the early 1880s, if not before (Hidden and Pratt, 1898).

The mining activity of the alluvial deposit progressed continuously from 1893 to 1901. The alluvial deposits of Mason's Branch were depleted by 1901, but there was some material found in mica schist or gneiss 'on the north side of the valley of Mason's Branch, especially near the summit of a knob where three openings were made' (Sterrett, 1911). The rhodolites were associated with biotite, quartz, pyrite, apatite, rutile, zircon and probably ilmenite, most of which are common inclusions in almandine (see inclusions in almandine in Chapter 4).

Ruby was apparently not found in the Mason's Branch, but only along the Cowee Creek deposits (Sinkankas, 1959). Inclusions in the garnets were biotite flakes and grains of quartz. When the alluvial material was finally depleted, exploration failed to reveal any other sources in the valley.

The year 1901 was a peak period for the rhodolite production. Approximately 200 000 carats of gem material were recovered during the year, having a value of $27 000 (Bauer, 1905; Kunz, 1903). But the yield for 1902 diminished considerably, to just $1500 (Sterrett, 1903). The total production figure for the nine year period was estimated by Kunz to have been approximately $53 000 (Sterrett, 1911).

One of the first operations along Cowee Creek occurred in 1895 when the American Prospecting and Mining Company of New York City started their commercial mining venture for the ruby along the creek (Sinkankas, 1959). The mining was quite sporadic over the next sixty or so years, but a revival of sorts began in the 1950s and 1960s when tourists came to the collecting sites, drawn by news of some striking ruby discoveries (Sinkankas, 1976).

The site at Mason's Mountain was reopened in 1961. Sinkankas (1976) reported that the site was exposed by an open pit dug into the hillside. Visitors were reportedly allowed to dig for a fee, but the difficulty with the rock and the techniques for mining the material made progress slow and arduous. Few pieces were recovered that were not shattered, although the mine owner was able to collect stones from time to time that would cut gems from ½ to 3½ carats (Sinkankas, 1976).

In the early period, mining along the creek bed at Mason's Branch was carried on with sluicing and hydraulic mining. The creek was even dammed up in order to facilitate the mining (Sterrett, 1911). The early material along the river was said to contain nearly all gem fragments, due to the agitation in the stream.

In this early period, 'rhodolite' seemed to find acceptance as a new 'variety'. Its colour was described in detail by Bauer (1905) as a

> 'pale rose-red inclining to purple like that of certain roses and rhododendrons . . . It lacks the depth and intensity of colour which makes garnets, as a rule, such dark-looking stones especially by artificial light. The particularly beautiful rose tint of rhodolite combined with its transparency and brilliancy renders it as even more striking object by candlelight than

by daylight. The lustre of rhodolite is comparable with that of demantoid, a green garnet from the Urals; this, together with its freedom from internal flaws and inclusions, makes it when cut a very striking and beautiful gem'.

Dr Bauer also mentioned that an examination of its spectrum by Professor Church revealed characteristic almandine absorption bands.

Rhodolite classification disputes

A few years later, 'new' sources of gem garnets were reported in North Carolina – in Macon and Jackson Counties. The gems were described as 'pink and purplish', some with practically the same colour as rhodolite, though generally of a dark shade' (Sterrett, 1911). The existence of these other garnets, some of which were identical to the original rhodolites, seemed to rule out colour as a distinguishing feature of the original material. Moreover, many later discoveries of nearly identical material seemed to undermine the uniqueness of the Macon County gems.

TABLE 5.1. Chemical analysis of rhodolite from Mason's Branch, North Carolina. The results are from a test performed by W. E. Ford, 1915

RI	1.7596
SG	3.837
Chemical analysis:	
SiO_2	41.59
Al_2O_3	23.13
Cr_2O_3	–
Fe_2O_3	1.90
FeO	15.55
MnO	–
CaO	0.92
MgO	17.23
End-member percentages:	
Pyrope	57.53
Almandine	36.77
Grossular	–
Spessartite	–
Andradite	5.70
Uvarovite	–

Furthermore, these new garnets did not represent new chemical end-members; consequently, mineralogists were reluctant to admit the name into the accepted garnet nomenclature. Objections to the name were also raised because 'rhodolite' sounded too much like rhodonite. However, to anyone familiar with rhodonite, there is little likelihood for confusing the two minerals. In a chemical analysis by Ford (1915) rhodolite proved to be 58% pyrope and 37% almandine (*Table 5.1*).

There were several attempts by later writers to accommodate the intermediate pyrope-almandine 'rhodolites' into the accepted garnet family. Some writers preferred to eradicate the intermediate category entirely and utilize the end-member names of pyrope and almandine only (Mackowsky, 1938). Others suggested the retention of an intermediate category, even if they objected to the 'rhodolite' name (Anderson, 1947; Anderson, 1959; Webster, 1983). Two problems exist with the latter option; one problem with the former. If the intermediary category is retained, then a name would have to be selected. Moreover, since there is a solid solution series between pyrope and almandine, the point selected to divide the species would be entirely arbitrary. The first option would have a similar problem; at what point along the RI/SG line would the division be made? Then would these divisions be suitable to classify future gem garnet deposits?

Anderson (1947) proposed the name 'pyrandine' (from *pyr*ope and alm*andine*) to represent the intermediate pyrope-almandine category. Webster called it the 'pyrope-almandine intermediate series' and suggested, after Anderson, an RI range of 1.75 to 1.78 and an SG range of 3.80 and 3.95 (Webster, 1983). Admittedly the divisions were arbitrary, but they did seem to divide the intermediate gems from known pyropes and known almandines of the period.

Another name suggested to represent the category was 'pyralmandite' by Fermor in 1926 (Fermor, 1948). In a study of the miscibility of garnets by Winchell and Winchell (1933) the name 'pyralspite' was used to describe the series, pyrope, almandine and spessartite. 'Pyralspite' continues to be widely used in mineralogy and petrology; but the name is not limited to just the pyrope-almandine category. It could also refer to spessartite-almandine garnets or even spessartite-pyrope intermediates. Moreover, the name does not imply a category between the end-members. Other names have also been proposed. Campbell (1972) suggested 'rhodomacon' in an attempt to associate rhodolite with its source. 'Umbalite' is another name proposed for intermediate garnets from Tanzania (as noted by Manson and Stockton, 1982). Both source names are quite inadequate, for they imply that the material is unique to the source.

If the intermediate category is to be retained as a special classification in gemmology, then there must be RI and/or SG limits for proper classification. Anderson's limits (1.75 to 1.78; 3.80–3.85) seem adequate enough; however, there is the problem of classifying stones with only one property within the range, whether it be RI or SG, and the other property out of the range. Hanneman (1983) saw this problem and suggested using only the RI range, deleting the SG range altogether. This may be the best alternative, even though the technique may oversimplify the process at the expense of the nature of the material.

There seems to be a dire need to study the relationships between known sources of these garnets and their essential gemmological properties, including colour variations, RI and SG values and chemistries (both bulk and trace chemistries). The studies of Manson and Stockton have contributed some answers to gem garnet classification and the role of colour in its identification. But the issues are far from being settled.

As it currently stands, 'rhodolite' is used largely as a trade term rather than an accepted mineralogical species or variety. The colour is tied to a violet-red position and exhibits an RI range of 1.75 to 1.78 and an SG range of 3.80 to 3.95. These values are remarkably adequate for most 'rhodolite' deposits and other intermediate garnets of the series, even though they are arbitrary divisions. Rhodolite, because of its almandine content, can be expected to reveal the typical almandine spectra. Inclusions also follow the almandine species.

Rhodolite sources

In addition to the North Carolina sources of rhodolite, long since depleted, there are many other reported worldwide deposits.

It is not known when the 'rhodolites' from North Carolina were first compared to the Ceylon gemstones. Certainly Hidden, Pratt and Bauer did not mention them in the late 1800s. However, a significant quantity of gemstones is produced there. These gems are called 'rhodolite' by the gem trade. The colour is violet-red and the RI/SG values are generally well within the intermediate category of Anderson and Webster, although the almandines from the area may also grade into the intermediates (Zwaan, 1967; Arbunies *et al.,* 1975).

'Rhodolites' have also been studied from Madagascar, although there is some doubt about their colour. Tisdall (1962) reported a Madagascar intermediate pyrope-almandine of purplish-red colour. Another study was conducted of ten Madagascar stones; their colour was also described as 'identical' to the North Carolina rhodolites (Campbell, 1973). But in a recent study of eighteen Madagascar gems the colour was analyzed using three systems of colour grading: the entire sample proved to be orange-red rather than violet-red (Rouse, 1984). RI and SG observations from all three studies placed the Madagascar garnets within the traditional intermediate range (RI: 1.749–1.769).

Rhodesia (Zimbabwe) has also been a source for rhodolite garnets. Campbell (1972) studied twelve stones and reported that they all ranged in colour from rose-red to pale violet. The RI range was from 1.750 to 1.760; the SG from 3.83 to 3.89, fitting well into Anderson/Webster intermediate values.

Another major source of rhodolite has been Orissa in India. Those studied with the Madagascar stones were 1.749 to 1.759 in RI; the SG was 3.73 to 3.86 (*see* Appendix 1). The SG values were generally below the cutoff recommended by Anderson and Webster; only five of the twelve stones were over 3.80 in SG. Moreover, they were largely violet-red in colour; yet, several were orange-red, revealing an inconsistency of colour in the source.

Greenland was also reported as a rhodolite producer, but it has not been a significant source (Trumper, 1952).

Perhaps the most important modern source for rhodolite is the prolific deposit in East Africa. The Umba valley rhodolites of Tanzania/Kenya border areas have been known since about 1964. Zwaan (1974) studied twenty-nine stones; their RIs ranged from 1.749 to 1.769 while their SG ranges were from 3.790 to 3.908, nearly identical to the Madagascar intermediates. Molecular end-member percentages (*Table 5.1*) indicated a dominant pyrope (60.27%) in the stone of the lowest properties (1.749; 3.79); but the stone with the highest values (1.769; 3.908) revealed a lower pyrope percentage (46.53%) with a nearly equal percentage of almandine (45.20%).

An earlier study by Bank and Nuber (1969) reported property values of the Tanzanian garnets to be somewhat lower than those studied by Zwaan. They varied between 1.745 to 1.755 with densities from 3.79 to 3.80.

TABLE 5.2. Molecular end-member percentages based on chemical analysis of Tanzanian rhodolites studied and reported by Zwaan (1974)

Sample	RGM 163–175	RGM 163–166	RGM 163–138	RGM 163–164
RI	1.749	1.756	1.762	1.769
SG	3.790	3.861	3.883	3.908
End-member molecule percentages:				
Almandine	21.52	35.76	38.00	45.20
Andradite	–	–	–	–
Grossular	7.61	5.30	8.33	4.97
Pyrope	60.27	57.62	48.00	46.53
Spessartite	10.60	1.32	5.67	3.30
Uvarovite	–	–	–	–

Two other important studies of rhodolite concentrated on analyzing properties and attempting to discover distinguishing characteristics of the rhodolite gems. Martin (1970) studied one stone from the original North Carolina source and eleven gems of 'rhodolite' colour, but of unknown sources. The North Carolina stone revealed an RI of 1.758 with characteristic almandine spectra; the eleven 'rhodolite' lookalikes proved to be pyrope, ranging from 1.745 to 1.750. The eleven test stones were suspected to be of Tanzanian origin, but inclusion studies and property analysis could not shed any light on either the origin or distinguishing features whereby they might be classified. The inclusions were quite typical of other almandine garnets, and nothing was found that might be unique to the source.

From many hundreds of East African rhodolites that we have studied, cut and collected, the sizes have been commonly up to 25 carats, occurring as river-tumbled pebbles. The clarities have been typically quite clean, although they can be included with rutile needles, sometimes in such profusion that the brilliance is affected. The colours range from violet-red to pale violet, with some percentage into the slightly orange-red hues.

Dealers specializing in East African gems have claimed that there is an unusually large abundance of rhodolite in the area, perhaps enough to last for decades. Rhodolite popularity in Japan created wholesale price increases several years ago (Osman, 1980). High demand internationally for these attractive stones has not abated.

Although 'rhodolite' as a gemmological term will probably remain in the vocabulary as long as there are gemstones which resemble the original North Carolina material, the term does not include all the gems to be found in the intermediate pyrope-almandine category. There are many gems which share some features of the rhodolites, especially their property values, but not their unique colour. Furthermore, there are some 'rhodolite' colours which exhibit property values outside the traditional boundaries of 1.75–1.78. For example, GIA reported two stones of typical rhodolite colour; one revealed an RI of 1.731, while the other was above 1.79 (Gems & Gemology, Summer, 1972).

Malaya garnet: a pyrope-spessartite intermediate

During the late 1970s, another garnet made its debut into the gemmological world.

For some years gem dealers in Nairobi would select stones from large parcels of rhodolite. Instead of the traditional violet-red hue, there were a number of other colours that did not fit the description of rhodolite. These ranged from orange to reddish-orange, generally with pinkish overtones. Their appearance was so obvious that they were sorted out as 'rejects' by the rhodolite buyers. The rejected stones were called 'malaya' which was the Swahili word for 'prostitute' (literally 'out of the family').

However, some dealers were fascinated with the gems and considered them very attractive in their own right. Finally one dealer submitted samples for testing. Some of the stones proved to be a combination of spessartite-pyrope, a mix of garnets not previously known to exist. Anderson (1959), for example, declared that 'manganese-pyropes do not seem to occur'.

Consequently, the gem dealers were convinced that their 'malaya' garnets were a new variety which deserved a new name. In 1979 and subsequent years there was a great deal of trade advertising displaying these 'new' garnets to the jewellers and gem buyers. Articles appeared in trade magazines describing their features and promoting their attractiveness. Brown (1981) described the colours as variable, ranging from gold to peach to orange to cinnamon to brown. Furthermore, their body colours 'were modified to varying degrees by the presence of a pinkish hue'. The source of these gemstones was identified to be 'near the village of Mwajijembe' in the Umba Valley of Tanzania; other sources were also reported across the border into Kenya (Curtis, 1980). However, the same species mixture (spessartite-pyrope) were also reported to be found also in Ratnapura, Sri Lanka (Schmetzer and Bank, 1981), so they did not seem to be unique to this source.

Another interesting feature of some 'malaya' stones was the presence of vanadium and chromium in varying amounts. Such stones with vanadium or chromium trace elements have been colour-change gems. Those stones with significant amounts of these trace elements did indeed produce such a change of colour. The colours were reported to change from magenta under incandescent light to bluish-mauve in fluorescence (Brown, 1981). However, future studies of colour change gemstones should be conducted under controlled, standardized grading lights, the reason being that the fluorescent lamp (standard office fluorescent) is a poor choice to approximate daylight, since it exaggerates the blues at the expense of the reds, quite unlike most daylight. Because of its unbalanced spectral emission distribution, one would expect some change of colour due to the lighting, rather than the gemstone.

In a major study involving 204 'malaya' garnets, chemical analysis revealed unusual end-member compositions, including grossular, spessartite and the unusual spessartite-pyrope intermediates (*Table 5.3*). The study was also extended to the colours, which were found to vary considerably (Jobbins *et al.*, 1978). Interestingly, some stones of this group proved to be pyrope-almandine intermediates, with little spessartite in the composition.

TABLE 5.3. Molecular end-member percentages based on chemical analysis of Tanzanian Umba Valley 'Malaya' garnets, from Jobbins, *et al.*, 1978

Ref. No.	1/15	9/27	1/3	CCG	11/1	1/26	9/14	10/40	10/42
RI	1.739	1.747	1.753	1.757	1.760	1.761	1.779	1.788	1.798
SG	3.745	3.650	3.823	3.816	3.855	3.863	3.995	4.049	4.141
End-member molecule percentage:									
Almandine	14.6	3.3	33.8	0.3	6.9	37.3	9.6	5.1	4.6
Andradite	1.6	8.0	–	3.2	–	–	3.5	3.8	1.7
Grossular	2.4	85.7	3.1	8.6	17.7	6.8	5.9	11.3	2.6
Pyrope	73.0	–	60.7	47.5	35.2	50.7	24.4	10.7	4.8
Spessartite	8.3	2.8	2.3	38.1	39.3	4.7	56.0	69.0	86.2
Uvarovite	0.1	–	0.2	–	0.1	0.3	–	–	0.03
Goldmanite	–	0.2	0.02	2.2	0.8	0.1	0.6	0.1	0.03

In a detailed study of the properties of the 'malaya' stones, Schmetzer and Bank (1981) reported that the gems studied could be divided into two types. In type I stones, there was considerable mixture of spessartite-pyrope without an almandine component. The type II stones, on the other hand, were spessartite-pyropes with varying amounts of almandine percentages. The type II stones were the ones similar to those found in Ratnapura, Sri Lanka. The RI and SG ranges fit well into those reported by Jobbins. It was recommended that, since the gem was not unique to known garnet deposits and since it presented no new end-member possibility, that no new name should be adopted. Furthermore, they proposed a name which was already in use: 'pyralspite' especially since the new stones were combinations of spessartite-almandine-pyrope, or pyrope-spessartites (Schmetzer and Bank, 1981).

Other pyrope-almandine intermediates ('pyrandine')

Excluding rhodolites, which are gems tied to a violet-red hue position, and the 'malaya' gems which present unusual colour and property mixtures, there are very many other gems which also must be considered in this category. The remaining intermediates may be almandines with low property values, or pyropes with high property values. Consequently, they may be found in either pyrope or almandine sources around the world.

Their colours tend to be orange to red in either medium or dark tones. The RI range is arbitrarily selected between 1.750 to 1.780, following Anderson and Webster's recommendations. Although gem rhodolites tend to cluster around 1.750–1.770, the off-colour rhodolites are broader in range (1.750–1.780), following the solid-solution series common in garnet minerals between pyrope and almandine.

Separation and classification

Gemmologically, the pyrope-almandine mixtures can be separated simply on the basis of the RI properties of the pyrope-almandine intermediate category (1.75–1.78). Colour analysis will serve to determine if the stones are 'rhodolite' intermediates, and absorption spectrum studies will properly satisfy requirements for 'malaya' intermediates (manganese spectrum). The stones that do not fit into the 'rhodolite' or 'malaya' categories can be simply be known as 'pyrandine' after B.W. Anderson's recommendations in 1947. The one flaw with this proposal is in the adoption of another gemstone to be separated on the basis of colour, similar to the ruby-pink sapphire and green beryl-emerald problems. Nevertheless, 'rhodolite' has always been tied to a definite hue position (violet-red), and most of the stones can easily be defined properly.

The mineralogical term 'pyralspite', although favoured by some mineralogists and petrologists, is too vague to apply to all three groups of intermediate garnets described above, except in a general sense.

It can also be seen that the above definition of 'malaya' garnet is made without reference to its source. The name could quite properly apply to any further deposits of spessartite-pyrope, as well as those currently found in East Africa and those in Sri Lanka as reported by Schmetzer and Bank (1981).

Intermediate pyrope-almandine and pyrope-spessartite properties

Gemmological classification:
'Pyralspite'
'Rhodolite'
'Malaya'
'Pyrandine'

Refractive index:	*1.750 to 1.780
Specific gravity:	3.80 to 3.95
Absorption spectra:	Rhodolite: Typical iron-rich pyrope: 575, 527 and 505 Malaya: May exhibit Mn lines at 412 and 432 Pyrandine: Same as rhodolite
Hardness:	7–7¼
Colours:	Rhodolite: Violet-red in various tones; usually lighter in tone than typical pyrope or almandine Malaya: Orange to orange-red. Pale tones not uncommon. Usually pink mixes with the basic hues, producing rather distinctive colours Pyrandine: Orange to orange-red. Usually medium to dark tones.

* The RI is the key for identification; these are the traditional values of Anderson and Webster.

Bibliography

ANDERSON, B. W., 'Pyrandine – a new name for an old garnet', *The Journal of Gemmology* (April, 1947)

ANDERSON, B. W., 'Properties and classification of individual garnets', *The Journal of Gemmology* (January, 1959)

ARBUNIES-ANDREU, M., BOSCH-FIGUEROA, J. M., FONT-ALTABA, M. and TRAVERIA-CROS, A., 'Physical and optical properties of garnets of gem quality', *Fortschr. Miner.*, **52,** Stuttgart (December, 1975)

BAUER, MAX, *Precious Stones,* Vermont and Tokyo, 1970. From the 1905 edition, trans. by L. J. Spencer

BROWN, GRAHAME, 'The malaya garnet', *Jeweller Watchmaker & Giftware* (March, 1981)

CAMPBELL, IAN C. C., 'A comparative study of Rhodesian rhodolite garnet in relation to other known data and a discussion in relation to a more acceptable name', *The Journal of Gemmology* (April, 1972)

CAMPBELL, IAN C. C., 'A gemological report on rhodolite garnet, Malagasy', *Lapidary Journal* (September, 1973)

CURTIS, COLIN M., 'Malaya-lady of the evening', *Lapidary Journal* (February, 1980)

FERMOR, L. L., 'Correspondence', *The Journal of Gemmology* (October, 1948)

HANNEMAN, W. WILLIAM, 'A new classification for red-to-violet garnets', *Gems & Gemology* (Spring, 1983)

HIDDEN, W. E. and PRATT, J. H., 'On rhodolite, a new variety of garnet', *American Journal of Science,* Vol. V (April, 1898)

JOBBINS, E. A., SAUL, J. M., STATHAM, PATRICIA M. and YOUNG, B. R., 'Studies of a gem garnet suite from the Umba River, Tanzania', *The Journal of Gemmology,* **XVI,** 3 (1978)

KUNZ, GEORGE F., 'Rhodolite', *Mineral Resources* (1903)

KUNZ, GEORGE F., *Gems and Precious Stones of North America,* New York (1890)

KUNZ, GEORGE F., 'Garnet', *Mineral Resources* (1899)

MACKOWSKY, MARIE-THERESE, 'The determination of garnets by means of their physical properties', *The Gemmologist* (September, 1938)

'Madagascar garnet production reported', *Gems & Gemology* (Winter, 1950–1)

MANSON, D. VINCENT and STOCKTON, CAROL M., 'Gem garnets in the red-to-violet color range', *Gems & Gemology* (Winter, 1981)

MANSON, D. VINCENT and STOCKTON, CAROL M., 'Gem garnets: the orange to red-orange color range', *International Gemological Symposium Proceedings,* 1982, ed. by Dianne M. Eash

MARTIN, B. F., 'A study of rhodolite garnet', *The Journal of Gemmology,* **XII,** No. 2 (April, 1970)

OSMAN, DANA, 'Malaya – the latest in a full spectrum of garnets', *Gems and Minerals* (January, 1980)

'Rhodolite Variations', *Gems & Gemology* (Summer, 1972)

SAROFIM, E., 'Gem rich Tanzania', *Lapidary Journal* (June, 1970)

SCHMETZER, KARL and BANK, HERMAN, 'Garnets from Umba Valley, Tanzania: is there a necessity for a new variety name?', *The Journal of Gemmology,* **XVII,** No. 8 (October, 1981)

STERRETT, DOUGLAS B., 'Garnet', *Mineral Resources* (1911)

TISDALL, F. S. H., 'Tests on Madagascar garnet', *The Gemmologist* (June, 1962)

TRUMPER, L. C., 'Rhodolite and the pyrope almandine series', *The Gemmologist* (February, 1952)

TRUMPER, L. C., 'Observations on garnet', *The Journal of Gemmology,* **VIII,** No. 8 (October, 1962)

WEBSTER, ROBERT, *Gems,* fourth edition (London, 1983)

ZWAAN, P. C., 'Some notes on the identification of the pyrope-almandine garnets', Koninks. Nederl. Akademie van Wetenschappen-Amsterdam. Reprinted from Proceedings, Seriel B. 64, No. 2 (1961)

ZWAAN, P. C., 'Solid inclusions in corundum and almandine garnet from Ceylon identified by x-ray powder photography', *The Journal of Gemmology,* **X,** no. 7 (1967)

ZWAAN, P. C., 'Garnet, corundum and other gem minerals from Umba, Tanzania', *Scripta Geol.,* **20** (1974)

Chapter 6

Spessartite

Unlike both pyrope and almandine, spessartite is very rare as a gemstone. There is no history of the gem before the 19th century; in fact, Mohs (1825) did not mention the gem, nor did Feuchtwanger in 1867. Even Bauer in 1896 simply made a passing reference to it in connection with the Amelia Courthouse deposit (Bauer, 1905). Miers (1902), however, mentioned spessartite, but failed to note the Amelia source, even though other sites were listed. Nor is it probable that these sources mistook it for hessonite, although there was some confusion between the two stones by some early writers (Sinkankas, 1959, p. 275).

Specific gravity tests were performed by serious mineralogists after Mohs' time and these would indicate that spessartite could not fit the hessonite description. Moreover, chemical analysis could also be used to separate hessonite and spessartite. Even in Mohs' time, chemical testing was performed which included the manganese content. What is most probable, is that the sources for the gemstone were simply not discovered, or that the few spessartites that might have come from known sources were simply unchecked and assumed to be other garnets.

Kunz was one writer who assumed the stone to be a variety of hessonite, since he described the Amelia gemstones as hessonites, 'in which the alumina is replaced by manganous oxide' (Kunz, 1890). Yet, in his chart for the various garnets, 'spessartite' was listed as a separate class of garnet, quoting sources from Philadelphia, Haddam, Connecticut and Nathrop, Colorado. The spessartites were described as being 'reddish, dark red, and dark reddish-brown' in colour, with specific gravities ranging from 4.12 to 4.27. Chemical analysis revealed that they were almandine-spessartite mixtures with the spessartite component dominant (over 50%).

Miers' sources for spessartite included the Haddam, Connecticut site and the Nathrop, Colorado deposit, but added a source in Belgium and Aschaffenburg in Bavaria (Miers, 1902). The material from Belgium was found 'in granules in the hone-stones of the Ardennes, and in a black metamorphic quartzite from Bastogne'. The Bavaria source was in the district of Spessart, which provided the original name to the species. But the major source of gem spessartite in the late 19th century was the Amelia site in Virginia. Even Bauer (1905) stressed that spessartite 'from other localities is scarcely suitable for cutting as a gem'.

Spessartite chemistry: $Mn_3 Al_2 (SiO_4)_3$

Spessartite was recognized very early to be a garnet rich in manganese. From Kunz' analyses, it is also obvious that the gem often contained considerable ferrous iron of the almandine component, although he listed a category called 'typical analysis' which gave a very high spessartite (over 90% Mn) with no ferrous iron. The varying percentages of iron in the spessartite indicated a solid solution existed between almandine and spessartite. The presence of magnesium in only trace amounts revealed that perhaps no solid solution existed between pyrope and spessartite until the discovery of the 'malaya' garnets of Tanzania in the late 1960s confirmed the solubility between the two end-members (Jobbins *et al.*, 1978).

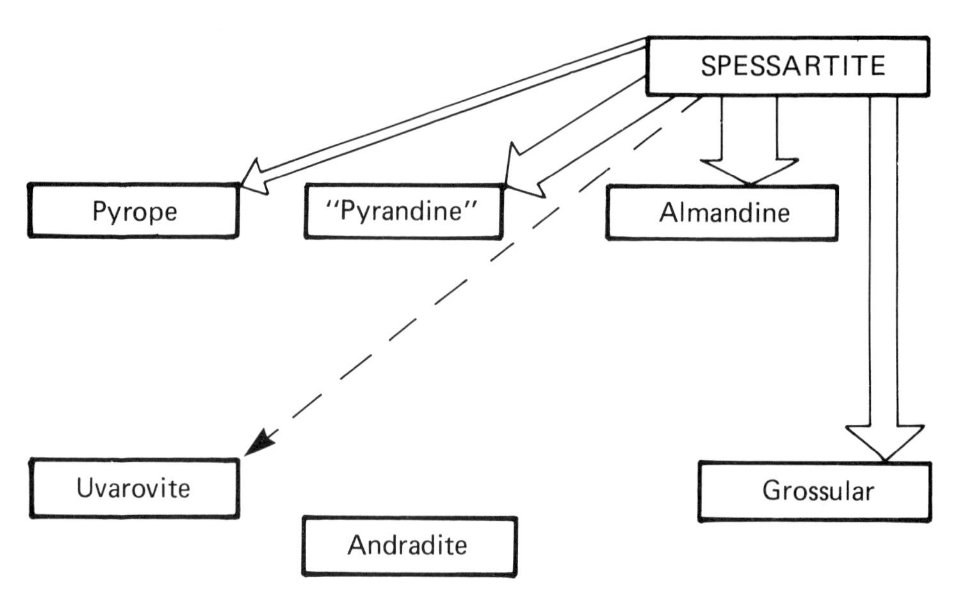

Figure 6.1 The miscibility relationship with other end-member garnets. Major arrows indicate probable or established solubility; dotted line indicates a possibility based upon an experiment with synthetic end-members (Naka *et al.*, 1975)

Also, in spite of early notions that only pyrope and almandine as well as spessartite and almandine were in solid solution, recent studies demonstrate that solid solution solubility probably extends across the 'pyralspite' boundaries into the 'ugrandite' series as well (see *Figure 6.1*). One study of synthetic uvarovite and synthetic spessartite revealed that a continuous solid solution was indeed possible at certain pressures and temperatures (Naka *et al.*, 1975). Moreover, continuous solid solubility can also be expected to occur between grossular and spessartite, since natural specimens have been found (Němec, 1967; Lee, 1962). Only future deposits of such mixtures will demonstrate whether they exist as gem material or mineral specimens.

From the spessartites exhibited in *Table 6.1*, it can be seen that most samples reveal no more than 12% grossular, with one sample reaching as high as 23.9%, perhaps reinforcing the solid solubility concepts advanced above. Other solid solution mixtures seen in *Table 6.1* are the almandine-spessartite and the pyrope-spessartite samples.

Spessartite colours

In the case of spessartite, it is apparent that there is a distinct relationship between the chemistry and the colours of the gemstones. A study of fifty spessartites from Amelia, Virginia indicated a colour range from pale orange to bright orange to a dark brownish-red (Sinkankas and Ried, 1966). The stones with high RIs and larger percentages of both grossular (to 7%) and almandine (to 43%) were generally in the dark reddish-brown colour range. Those stones with smaller percentages of grossular (2–3%) and almandine (2–20%) tended to exhibit colours in the pale orange hues. Stones of the latter group also exhibited the lowest RI properties. The authors speculated that a pure spessartite would probably be 'extremely pale orange or possibly colourless'.

Spessartite colours are variously described as yellowish-orange to 'aurora-red' to brownish-red, depending on the source and chemical mixture. Since the 'aurora-red' colour is often described as the finest hue position, there must be some almandine present to alter the hue. Moreover, the colours of some spessartites overlap those of the grossular, in which case identification by refractive index or specific gravity testing will serve to separate them. Colour alone should not be relied upon, therefore, for sight identification of spessartite; gemmological testing must be resorted to.

The tones of the spessartite hues can range from very pale to very dark. Fine, vivid intensities tend to be found in the mid-tonal ranges ('aurora-red'), but they are very rare and highly prized when found. Very dark spessartites (iron-rich) tend to exhibit weak to dull intensities. Pale hues are also weak in intensity, but this condition is due to the lack of colour in the sample.

The colour in spessartite is due to a mixture of manganese and iron in the sample. But there may be trace elements which contribute to the colour as well. There also may be instability in the colour of some purish orange spessartites, as one report indicated some fading when the stone was exposed to sunlight (Smith, 1977, p. 336).

Optical and physical properties

Spessartite is considered to be over 7 in hardness, although Kunz reported that the 'typical' spessartite exhibited a hardness of 6.5 (Kunz, 1890, p. 84). Webster places the hardness at 7¼ (Webster, 1983), which is generally accepted, however.

TABLE 6.1. Spessartite garnets reported in the literature, giving optical, density and calculated end-member components

Specimen	1	2	3	4	5	6	7	8	9	10
RI	1.779	1.788	1.790	1.795	1.796	1.798	1.798	1.798	1.805	1.809
SG	3.995	4.049	4.07	–	–	4.15	4.14	4.141	4.117	–
Major molecule end-members (calculated)										
Spessartite	56.0	69.0	70.7	95	90.03	90.30	60.7	86.2	61.6	50
Pyrope	24.4	10.7	13.4	–	0.79	0.90	5.4	4.8	–	–
Grossular	5.9	11.3	7.4	2	4.82	7.90	23.9	2.6	5.5	7
Andradite	3.5	3.8	5.0	–	–	0.20	1.5	1.7	–	–
Almandine	9.6	5.1	2.8	3	4.36	0.70	8.4	4.6	32.9	43
Calderite	–	–	–	–	–	–	–	–	–	–
Schloromite	–	–	0.60	–	–	–	–	–	–	–
Hydrogrossular	–	–	0.10	–	–	–	–	–	–	–
Goldmanite	0.60	0.1	–	–	–	–	–	0.30	–	–
Uvarovite	–	–	–	–	–	–	–	0.30	–	–

1. Alluvial, Umba Valley, Tanzania; Jobbins, *et al.* (1978)
2. Alluvial, Umba Valley, Tanzania; Jobbins, *et al.* (1978)
3. In pyrrhotite skarn rock, Meldon, Okehampton, Devonshire; Deer, Howie and Zussman (1962) recalculated by Rickwood (1968)
4. Pegmatite, Rutherford mines, Amelia, Virginia; Sinkankas (1966)
5. Manganese mineral vein in pegmatite, Bald Knob, North Carolina; Ross and Kerr (1932)
6. In schist, Arrow Valley, Kawarau Syruvey District, Western Otaga, New Zealand; Deer, Howie and Zussman (1962) recalculated by Rickwood (1968)
7. Metachert associated with glaucophane schists, Valley Ford, California; Deer, Howie and Zussman (1962) recalculated by Rickwood (1968)
8. Alluvial, Umba Valley, Tanzania; Jobbins, *et al.* (1978)
9. Pegmatite; Avondale, Pennsylvania; Deer, Howie and Zussman (1962) recalculated by Rickwood (1968)
10. Pegmatite; Rutherford mines, Amelia, Virginia; Sinkankas (1966)

The refractive indices and specific gravity values overlap almandine. The RIs of the Amelia spessartites ranged, rather narrowly, from 1.795 to 1.809 (Sinkankas and Reid, 1966). Specific gravity tests were not conducted on the fifty Amelia stones, however. One of the earliest studies on the Amelia spessartite revealed an SG of 4.20 on a sample analyzed (Fontaine, 1883). Other samples (*Table 6.1*) reveal specific gravity ranges from 3.995 to 4.15, in some contradiction to Webster's limits of 4.12–4.20 (Webster, 1983). Another study of almandine-spessartites from unknown sources indicated a range from 4.13 to 4.33 (Manson and Stockton, 1981).

The latter study also indicated a range of RI between 1.794 and 1.820, as opposed to Webster's 1.79–1.81 range (Webster, 1983). Because of this overlapping with almandine, refractive index and specific gravity tests will not be useful for separating the two gemstones. Only spectroscopic analysis and perhaps inclusions studies are useful to separate almandine from spessartite.

Pure spessartite ideally exhibits RI/SG values of 1.800 and 4.19 respectively (Skinner, 1956). Although spessartites in nature can be found close to these ideals, pure spessartite has not yet been found to exist in nature. Studies of the Amelia gemstones revealed spessartite components up to 96% Mn. However, the purest specimens were reaching just below Skinner's ideal value of 1.800 (Sinkankas and Reid, 1966). It is also interesting to note that in the Amelia spessartites, the ratio of almandine and calcium follows a similar pattern: both diminish in percentage as the RI diminishes.

Except for the first two stones of *Table 6.1*, the others fit the traditional RI values of 1.79–1.81 quite well. However, it must be understood that chemical composition determines the precise identity of the garnet; in the case of spessartite, most of the stones can be identified readily by RI, SG, colour, and absorption spectra. However, low property spessartites may remain a problem for gemmologists, for only chemical analysis can confirm the identity with any degree of certainty.

Absorption spectra

Fortunately for gemmologists, spessartite gemstones exhibit a characteristic spectrum. Otherwise, separations between almandine and spessartite, in particular, would be difficult.

Most of the characteristic bands of spessartite occur in the dark end of the spectrum, where lighting is difficult for observation. However, a strong characteristic band often occurs as a cuttoff at 432 nm. If the beginning of the band is not visible, certainly the end of the band is quite evident.

For those students whose eyes are sensitive to the violet end of the spectrum, another strong band is found at 412 nm, with an accompanying fine line at 424 nm. Other lines sometimes seen are a weak line at 462 and weak fine lines at 485 and 495. The latter lines must not be confused with the iron bands sometimes found in the same region, at 476, 462 and 438 (Anderson and Payne, 1955, 1956).

Liddicoat (1972) found three additional lines in a light orange spessartite from Amelia. These occur as weak, narrow bands at 510, 532 and 560. These latter lines are situated quite close to the traditional iron bands at 505, 525 and 575. However, in the spectra of the Amelia spessartite, no iron absorption bands were seen in the sample.

Like the almandine spectra, the strength of the bands and lines will vary depending on the percentage of manganese present in the sample. In high Mn samples, the lines will be more distinct; in lower percentates of Mn, the pattern is weaker.

Moreover, since the almandine forms a solid solution series with spessartite, iron is a common ingredient, and the spectra will reveal the iron bands. But even in almandine-spessartite intermediates, the strength of the iron spectra will reveal the percentage of iron in the sample; however, most students cannot distinguish this subtlety without considerable practice with many samples.

Spessartite inclusions

Although spessartite is unusually free of internal flaws, there are several types of inclusions to be found. It is quite common to find liquid-filled inclusions in many forms. Occasionally solid crystals and two-phase inclusions (liquid and gas) can be seen; negative crystals are quite common.

In some instances the negative crystals and the liquid droplets form a pattern not unlike a fingerprint. But the characteristic display of inclusions in spessartite can best be described as 'wavy feathers formed by minute liquid drops of peculiarly shredded appearance . . . These flaws either are embedded singly in the spessartite, or else combined in bundles, and extend in every possible direction' (Gübelin, 1953).

In the Amelia spessartites, it was found that colour was also a factor in the amount of inclusions. The pale orange colours seemed to produce the most inclusions, while the darker colours (higher percentate of iron from the almandine) seemed to be less included (Sinkankas and Reid, 1966).

Spessartite sizes

Gem spessartites can occur in large sizes. However, the large sizes are rarely, if ever, seen in the commercial gem marketplace. The gem is so rare that only with new finds producing large quantities of material, is it possible to see even 5–10 carat cut samples. As the mines are depleted, smaller goods inevitably are found on the market.

Unlike almandine, which can be found in 75 mm and 100 mm diameter specimens in some quantities, spessartite is very rarely found in those dimensions. However, when spessartite is found in large sizes, it can be quite easily cut, especially since the tones are not nearly as dark as the almandine. Therefore, large cut spessartites will carry a price premium, simply because they retain elements of their beauty when cut into such large specimens. As a result, collectors will search out large cut gems which may sometimes exceed 100 carats in size.

The largest rough specimen of spessartite still appears to be the piece found by Sean Sweeney in September, 1972, from the dumps of the No. 2 Rutherford mine in Amelia, Virginia. It weighed 1.344 kg (6720 carats) and measured 5 × 4½ × 2¾ in (Sinkankas, 1976).

Another group of large stones was found in 1976 by John Nygaard (Sinkankas, correspondence, 1976). The largest was a piece weighing 1675 carats. Others in the same find weighed 717.5 carats, 385 carats, 285 carats, 37 carats and 17 carats. They were of gem quality with colours ranging from orange to red to 'wine-red'. A number of museums own specimens from about 5 carats to 40 carats in size.

The shape of the crystals is unusual for garnet. Because of complex crystallization, they are sometimes found in undefined masses with etched striations on the surface. They have occurred in this fashion in both the Amelia deposits and the Ramona sites (Sinkankas, 1959, p. 282).

Spessartite occurrence

One of the most common occurrences of spessartite is in pegmatites and in such a site, the spessartite is commonly a spessartite-almandine combination (Deer, Howie and Zussman, 1962). Also, the yttrium-spessartites seem to be restricted to this paragenesis (Deer, Howie and Zussman, 1962). Other occurrences include some skarn deposits, some manganese-rich assemblages, in gneiss, and in metamorphosed chert.

Spessartite sources

The earliest source of spessartite from Aschaffenburg (the Spessart district of Bavaria) produced yellow to red trapezohedron crystals in granite (Miers, 1902). However, there seems to have been very little material from the site, for few writers mention the deposit.

Certainly the most famous mines were the Rutherford Mines of Amelia Courthouse, Virginia USA. There were just two pits mined from the site known as the Rutherford No. 1 and Rutherford No. 2 mines. They were granitic pegmatites which produced a vast array of minerals in addition to the spessartite. The first commercial activity in the site started in 1873, even though about 3 m of material had already been removed by early settlers and Indians (Glass, 1935). The No. 1 mine was situated on the hill, while the No. 2 mine was in the creek bed.

Although there were other dikes in the area (Pegau, 1928), the two Rutherford pits provided most of the rare minerals from the district. The mines are located just 2.4 km north of Amelia (*Figure 6.2*).

The mining of the pegmatites proceeded until 1912, when the No. 2 mine was abandoned. The No. 1 mine continued to produce gem grade amazonite and continued to be worked until 1932. In 1943, the No. 2 mine was reopened for mica, but only for a short time. Both mines were reopened from 1957 to 1960 and were

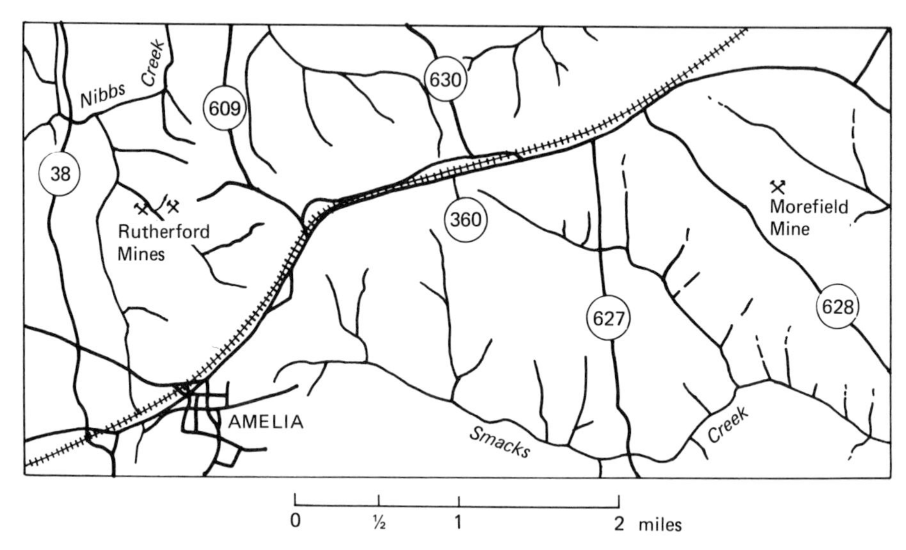

Figure 6.2 Location of the pegmatites near Amelia Court House, Virginia, USA (from Sinkankas and Reid, 1966)

systematically explored for old and new material (Sinkankas, 1968). When they were finally closed in 1960, the No. 2 pegmatite appeared to be pinched off at a depth of 35 m and was considered depleted (Sinkankas, 1968).

Although the spessartite and other minerals from the Rutherford mines were analyzed and described as early as 1883 (Fontaine), and briefly mentioned by many later writers, notably Pegau (1928) and Glass (1935), the study by Sinkankas (1968) is particularly revealing in connection with the spessartite from this source.

Spessartite crystals from the early workings of the Rutherford deposits produced poorly formed crystals which were penetrated with fissures and were quite easily shattered. They were also described as pale pink to 'flesh-red' with some fragments brownish-purple (Fontaine, 1883); however, they appeared more like rhodonite to that writer than garnet. Although the purplish coloured garnets were reported by Fontaine in 1883 ('brownish-purple') and Glass in 1935 ('deep wine-red'), the fifty samples studied by Sinkankas and Reid (1966) failed to show any purplish colour, which is considered to be associated with the almandine.

The crystals from the No. 2 mine have been known for their large sizes and their gemmy nature. Although the large crystals were generally quite flawed, they contained areas which were able to be cut. The profusion of veil-like inclusions (healed fractures) and fissures in the spessartite seems to suggest that massive cracking occurred during crystal growth, but because the crystals were physically restrained from movement, they tended to 'heal' back together (Sinkankas, 1968, pp. 401–402). Also reported by the same writer were ball-like aggregates of spessartite crystals which exhibited a characteristic zoning:

> 'the exterior crystals are orange-zoned only adjacent to the periphery of the aggregate itself. The general shape and color-zoning suggest that each ball-like aggregate was originally a single crystal which fractured late in its growth after it had accumulated an outer zone of orange-colored material'.

1.

2.

3.

4.

1. Group photograph of garnets, including, from the top centre, red rhodolite, medium red rhodolite, hessonite, malaya, green grossular, colourless grossular, pale yellow grossular, and pinkish-red rhodolite. Colour variations in rhodolite are quite obvious, as well as overlapping colours possible in malaya and hessonite
2. Medium orange malaya garnet from East Africa. Many colour variations exist in malaya. This medium orange is one of the most characteristic of the malaya colours
3. Very fine pinkish-red rhodolite from East Africa. The beauty of the hue position is matched by the striking intensity and the light tone
4. Tsavorite garnet in a medium green colour and fancy cut. The green hue is into the yellowish-green, but the brilliance and the medium tone combine to produce a fine gem.

Plate 1 Pyrope garnet from Orissa (RI 1.749). Needles of various length with disc-like platelets and tiny crystals. 57x (*WM*)

Plate 2 Pyrope garnet from Orissa (RI 1.758). Disc-like inclusions arranged in a row. 57x (*WM*)

Plate 3 Pyrope garnet from Orissa. Sparse needles in one direction. 50x (*WM*)

Plate 4 Intermediate pyrope-almandine from Malagasy Republic. Intersecting needles are displayed, a characteristic inclusion for almandine garnet and intermediates (pyrope-almandine). 45x (*JR*)

Plate 5 Garnet with a crystal inclusion from Malagasy Republic typical of the gem. 67x (*WM*)

Plate 6 Interesting twinned crystal in Malagasy Republic garnet. 50x(*WM*)

Plate 7 Partially resorbed crystal (?) in Malagasy Republic garnet. 100x (*WM*)

Plate 8 Unusual fingerprint inclusion found in a Malagasy Republic garnet. 25x (*WM*)

Plate 9 Hessonite from Malagasy Republic revealing large fissure-like inclusion, viewed under crossed polars. 45x (*WM*)

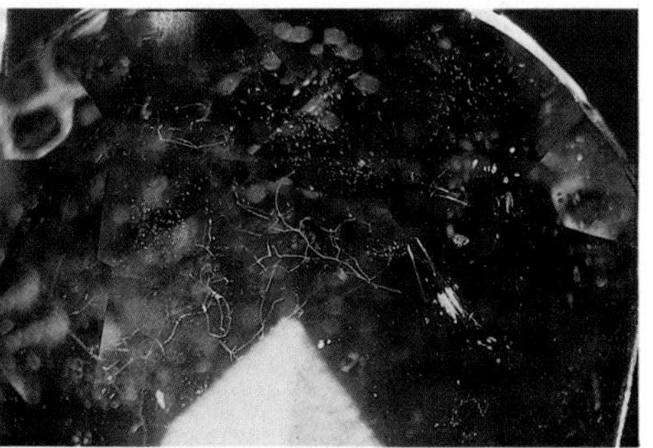

Plate 10 Unusual thread-like inclusions in hessonite garnet, believed to be from Baja, Mexico. 30x (*WM*)

Plate 11 Needle-like inclusions from Mexican hessonite. 32.5x (*WM*)

Plate 12 Interesting pattern of liquid and/or negative inclusions in Mexican hessonite. 35x (*WM*)

Plate 13 Two-phase inclusion (liquid, gas) in Mexican hessonite. 87.5x (*WM*)

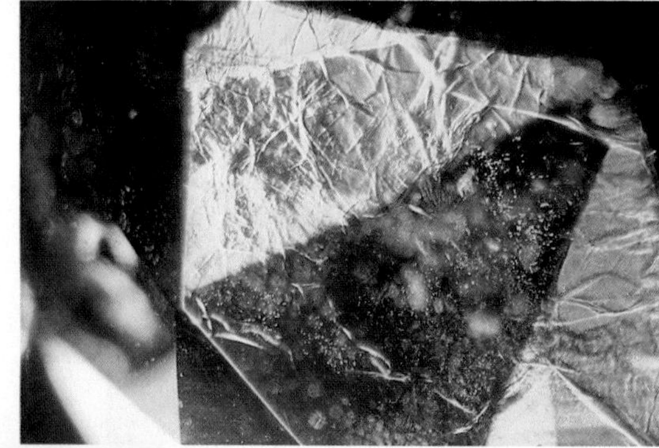

Plate 14 Heat-wave effect, a common occurrence in hessonite; Malagasy Republic hessonite, 42.4x (*WM*)

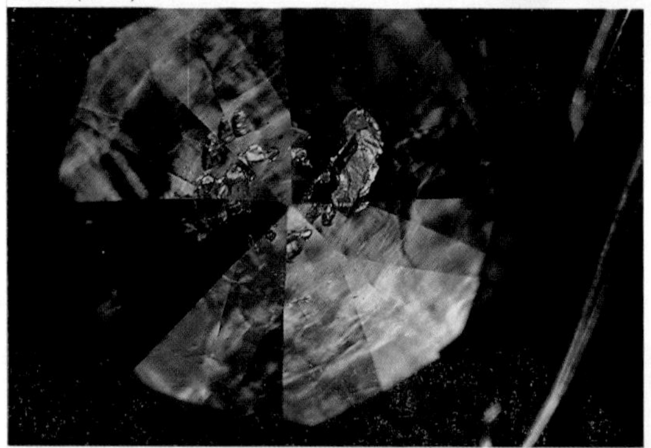

Plate 15 Strain, in association with heat-wave effect and crystal inclusions, seen in a hessonite from Malagasy Republic. 17.5x (*WM*)

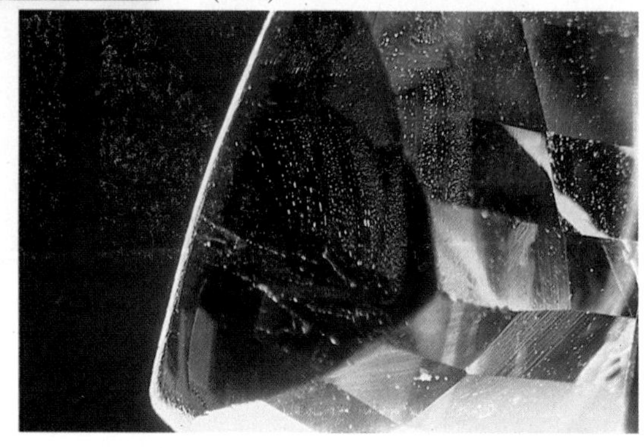

Plate 16 Fingerprint-type pattern in spessartite garnet of unknown source. Liquid and negative crystals make up the fingerprint (*WM*)

Plate 17 Liquid-filled cavities in spessartite garnet of unknown source. 82.5x (*WM*)

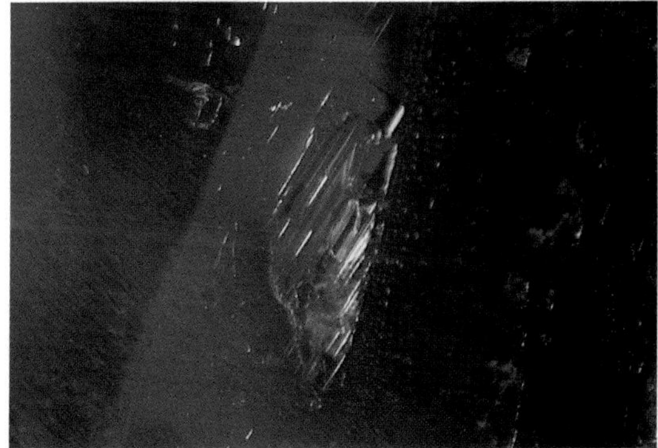

Plate 18 Interesting liquid-filled fracture in spessartite garnet of an unknown source (possible crystallizing fracture). 105x (*WM*)

Plate 19 Liquid-filled inclusions forming several fingerprint patterns in spessartite said to be from Amelia, Virginia. 30x (*WM*)

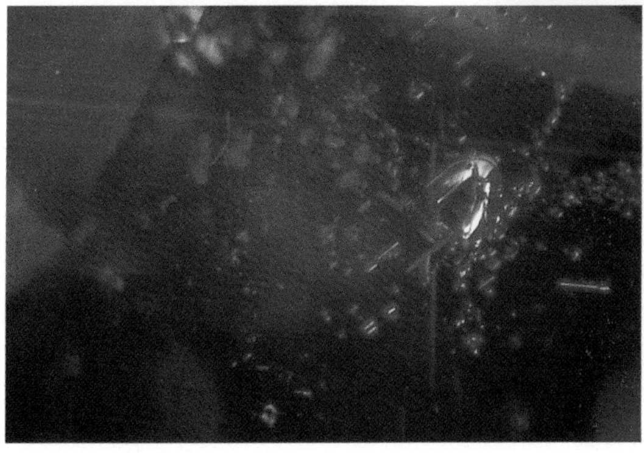

Plate 20 Rhodolite garnet from Ceylon revealing crystal with slight halo and several needles around it. 60x (*WM*)

Plate 21 Star garnet rhodolite from East Africa. By transmitted light from under the stone, a six-ray star occurs (*top left*); in reflected light from above the stone, a 4-ray star appears (*top right*); in transmitted light from the side of the stone, another 6-ray star can be seen (*bottom left*), representing a very rare phenomenon. 15x, 10.5x, 10.5x (*WM*)

Plate 22 An illustration of dense silk in rhodolite garnet. When this condition occurs, light entering the stone is diffused and the tone darkens as a result. 60x (*WM*)

Plate 23 Crystal plate-like inclusion in pale orange grossular garnet from unknown source. 60x (*WM*)

Plate 24 Same inclusion as #23 with crossed polars. 22.5x (*WM*)

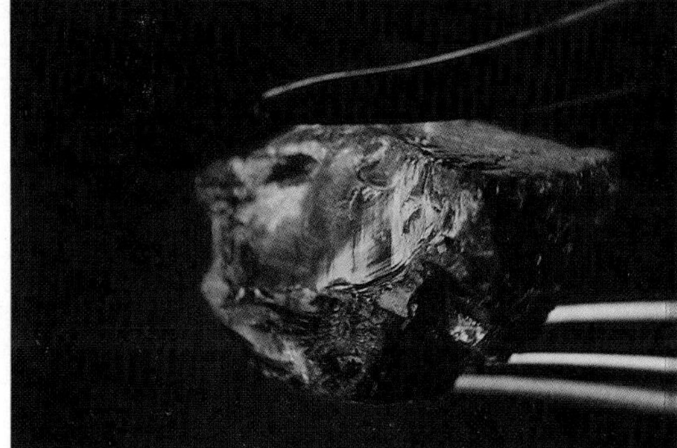

Plate 25 Spectral colour seen in rough gem grossular due to crystal growth. 10.5x (*WM*)

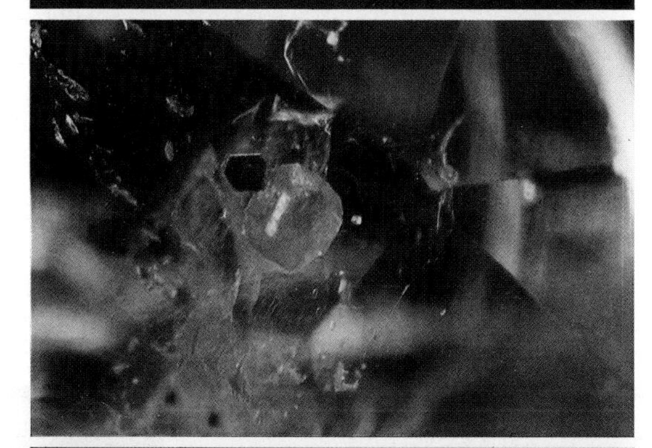

Plate 26 Crystals in pale orange grossular garnet from unknown source. 87.5x (*WM*)

Plate 27 Crystal cluster in tsavolite garnet from East Africa. 105x (*VR*)

Plate 28 Encrusted coarse needles (diopside?) in tsavolite from East Africa. 30x (*JR*)

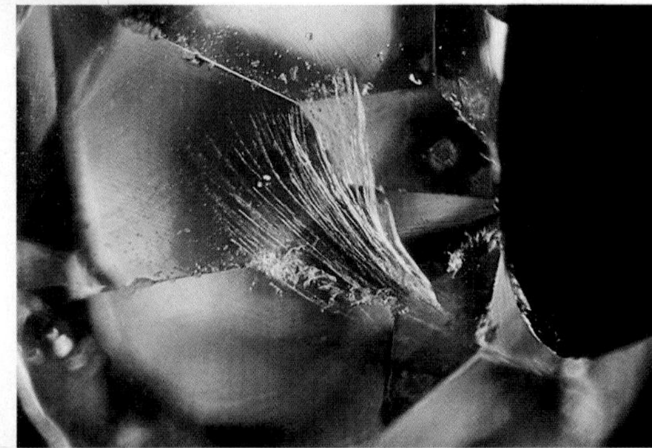

Plate 29 'Horsetail' byssolite inclusion typical of demantoid garnet. From andradite of unknown source. 57.5x (*WM*)

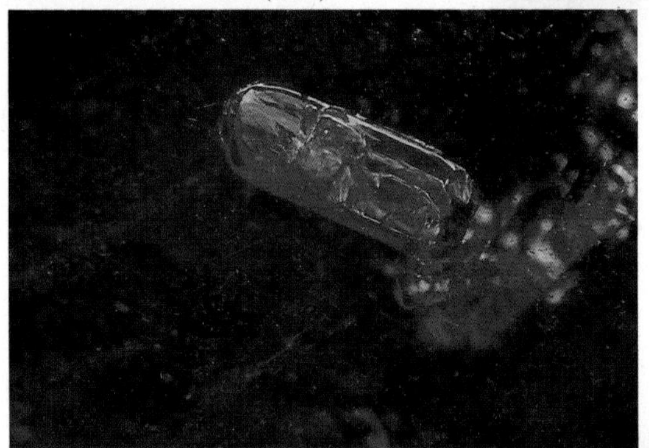

Plate 30 Possible apatite crystal in almandine garnet from unknown source. 45x (*WM*)

Plate 31 Crystal which surfaced on the table of an almandine from an unknown source. 45x (*WM*)

Acknowledgements

Plates 1–3, 5–26, 29–31. (*WM*) Wimon Manorotkul, lab gemologist, AIGS
Plate 27. (*VR*) Valaya Rangsit, GG, lab gemologist and instructor, AIGS
Plates 4, 28. (*JR*) Author

In addition to the spessartite of the Rutherford mines, the Morehead mine, situated nearly 6.4 km east of the Rutherford deposits (see *Figure 6.2*), also produced spessartite. The crystals here occurred in biotite schist and they were imperfect in shape. They were also coated with a black manganese oxide. Once the coating was removed by acid treatment, the spessartite colour was described as 'flesh-red' with an RI of 1.804. A spessartite-almandine mixture was also produced, which exhibited a deep red colour and an RI of 1.810 (Glass, 1935, p. 758).

Another important source for spessartite has long been reported to be the mines in the Ramona district of San Diego County, California. The Surprise and Hercules mines were reported in the 1905 annals of *The Mineral Industry* as active producers of spessartite. These mines were approx 7 km northeast of Ramona and the garnets occurred in pegmatite veins. The gems from the Hercules mine produced flawless stones up to eight carats in weight; these gems sold for $20 per carat, a high price considering the value of the dollar for 1905.

In more recent years the site has produced spessartite with colours described as very fine, delicate yellowish-orange of bright intensities (Tisdall, 1962, p. 124). Other mines have also been added to the original Surprise and Hercules sites. The Little Three mine was an extension of the Hercules mine and the deposit has produced gem quality spessartite. The Spaulding mine has also been a fruitful source of gem spessartite, in addition to the mines mentioned above. All of these deposits are very close together situated on a hillside and stretching down to Hatfield Creek.

The crystals of spessartite in these pegmatites have been found in trapezohedral forms up to several inches in diameter. The large specimens, however, are typically filled with cracks which limit the gem size to below 7 carats (Sinkankas, 1959, p. 287). Some of the crystal faces were reportedly smooth, but typically, many showed the same etched striations visible in the Amelia spessartites. The Ramona spessartite mines are still active producers of gem material.

The Pala District of San Diego County has also yielded some gem spessartite. Early workings of the Himalaya Mine and other Pala deposits have produced fine specimens, although there have been isolated occurrences in the pegmatites (Sinkankas, 1959).

In recent years, spessartite from the Malagasy Republic has produced significant quantities of gem material. The stones have been reported as brownish-red to a rich orange-red in colour; some have been quite dark and others pale, but they could not match the fine colours of the Ramona spessartites (Tisdall, 1962). The Malagasy spessartites are usually very free of internal inclusions, but the frequency of stones with a brownish-orange cast of the almandine component makes them somewhat less desirable than others from various sources. Tisdall's test of the Madagascar spessartite revealed an RI of 1.810 and an SG of 4.17; the spectra displayed a strong absorption in the violet, so that only the 432 nm band was seen as a cut-off ending at 442 nm (Tisdall, 1962). They also revealed the iron lines of the almandine mixture at 505, 527 and 575.

India and Sri Lanka are also producers of spessartite. In India spessartite has been reported in Narukot, Bombay. They were of a fine orange-red colour in some quantity (Iyer, 1961). Spessartite is also reported in the manganese mines of

Madhya Pradesh; the colours range from orange to brownish-red, but they are often heavily flawed (Iyer, 1961). Fermor (1938) also claimed that spessartite was abundant in India and it had been found in many places.

In a study of an unusual spessartite from Madhya Pradesh, chemical analysis revealed a calderite-spessartite. It contained a molecular percentage of 75.72% spessartite, 17.85% calderite, 5.93% andradite and just 0.50% of almandine. Calderite is a manganese ferric iron silicate $(Mn_3Fe_2(SiO_4)_3$ representing a rare garnet species. The sample tested was a coarsely crystalline orange-yellow aggregate. In thin sections the garnet was yellow and revealed inclusions of 'microcline, sodic plagioclase, biotite, quartz and ore' (Sastri, 1963).

Spessartite also is found in Burma, Norway, New South Wales, Australia, and Brazil. In Brazil gem material is found in the gravels of the Santa Maria and Abaete rivers as well as Arassuahy, Registo and Ceara (Webster, 1983).

A number of isolated early occurrences have been reported in the USA. Sterrett (1908) reported a source of spessartite on Ruby Mountain, on the east side of the Arkansas River just opposite Nathrop, Colorado. The spessartite occurred in rhyolite and was a transparent deep-red to cinnamon-red colour. The stones were small, measuring only 2.5 mm on the average.

Another source of spessartite, quite well-known to collectors, is from Avondale, Pennsylvania. The garnets occur in a pegmatite vein that cuts through a granite gneiss. Crystals occur in large sizes which have been collected over a period of many years. A sample has been included in *Table 6.1* (No. 9), which reveals a large percentage of iron (32.9%) in the chemical mix (Strock, 1930).

North Carolina also has produced spessartite in the vein of Bald Knob. Massive spessartite and fine-grained granular material is reported in the manganese-rich vein. The massive spessartite forms in lenses from 15 to 20 cm wide and produces an RI of 1.802 or 1.796 for the fine-grained granular. Chemical analysis revealed a high percentage of manganese (90.03%) and a low percentage of iron (4.36%) from this site (Ross and Kerr, 1932).

Spessartites and intermediate almandine-spessartite garnets have also been studied from pegmatites in Eilat in southern Israel (Nathan *et al.*, 1965). The crystals were small in size (½–1 mm), but transparent with a 'pinkish-brown' colour. One pegmatite was exceptional, in that it produced stones up to 3 cm in diameter.

Summary of spessartite properties

Chemistry:	$Mn_3Al_2(SiO_4)_3$
Colouring agents:	Manganese; iron
Refractive index:	1.79–1.81
Specific gravity:	4.12–4.20
Absorption spectra:	432 nm band, very strong, appearing as a cutoff
	424 nm
	412 nm intense band
	Also: weak bands at 495 nm, 485 nm, 462 nm
Hardness:	7¼
Colours:	Orange, orange-red, red, brownish-red
Dispersion:	0.027

Bibliography

ANDERSON, B. W., 'Properties and classification of individual garnets', *The Journal of Gemmology,* **7,** No. 1 (January, 1959)

ANDERSON, B. W. and PAYNE, C. J., 'The spectroscope and its application to gemmology, part 19: absorption of almandine garnet', *The Gemmologist* (March, 1955)

ANDERSON, B. W. and PAYNE, C. J., 'The spectroscope and its application to gemmology, part 25: concluding summary of absorption spectrum, *The Gemmologist* (December, 1956)

BAUER, MAX, *Precious Stones,* translated by L. J. Spencer, Vermont and Tokyo (1970 reprint of the 1905 edition)

DEER, N. A., HOWIE, R. A. and ZUSSMAN, J., *Rock-Forming Minerals. Ortho- and ring silicates,* **1,** London (1962)

FERMOR, L. L., *Garnets and their Role in Nature,* Calcutta (1938)

FEUCHTWANGER, L., *A Popular Treatise on Gems,* New York (1867)

FONTAINE, W. F., 'Notes on the occurrence of certain minerals in Amelia County, Virginia, *American Journal of Science,* 3rd Series, **125** (1883)

'Garnet', *The Mineral Industry,* **XIV** (1906)

GLASS, JEWELL J., 'The pegmatite minerals from near Amelia Virginia', *The American Mineralogist,* **20,** 11 (November, 1935)

GUBELIN, EDWARD J., *Inclusions as a Means of Gemstone Identification,* Los Angeles (1953)

IYER, L. A. N., 'Indian Precious Stones', Bulletin No. 18, *Geological Survey of India* (1961)

JOBBINS, E. A., SAUL, J. M., STATHAM, PATRICIA M. and YOUNG, B. R., 'Studies of a gem garnet suite from the Umba River, Tanzania', *The Journal of Gemmology,* **XVI,** No. 3 (1978)

KUNZ, GEORGE FREDERICK, *Gems and Precious Stones of North America,* New York (1890)

LEE, DONALD E., 'Grossularite-Spessartite garnet from the Victory Mine Gabbs, Nevada, *The American Mineralogist,* **47** (January-February, 1963)

LEWIS, DAVID, *Practical Gem Testing,* London (1977)

LIDDICOAT, R. T., *Handbook of Gem Identification,* Los Angeles (1972)

MANSON, V. and STOCKTON, CAROL, 'Gem garnets: the orange to red-orange color range', *International Gemological Symposium Proceedings,* ed. by Dianne M. Eash (1982)

MIERS, HENRY A., *Mineralogy* (1902)

MOHS, FREDERICK, *Treatise on Mineralogy,* 3 volumes, London (1825)

NAKA, SHIGEHARU, SUWA, YOSHIKO and KAMEYAMA, TETSUYA, 'Solid solubility between uvarovite and spessartite', *American Mineralogist,* **60** (1975)

NATHAN, Y., KATZ, A and EYAL, M. 'Garnets from the Eilat Area, Southern Israel', *Mineralogical Magazine,* **35,** 270 (1965)

NEMEC, D., 'The miscibility of the pyralspite and grandite molecules in garnets', *Mineralogical Magazine,* **37** (1967)

PEGAU, A. A., 'The Rutherford Mines, Amelia County, Virginia', *The American Mineralogist,* **13** (1928)

RICKWOOD, P. C., 'On recasting analyses of garnet into end-member molecules', *Contributions to Mineralogy and Petrology,* **18** (1968)

ROSS, CLARENCE S. and KERR, PAUL F., 'The manganese minerals of a vein near Bald Knob, North Carolina', *The American Mineralogist,* **17,** p. 1 (1932)

SASTRI, G. G. K., 'Note on a chrome and two manganese garnets from India', *Mineralogical Magazine,* **33** (June, 1963)

SINKANKAS, JOHN, 'Classic mineral occurrences: I. Geology and Mineralogy of the Rutherford pegmatites, Amelia, Virginia', *The American Mineralogist,* **53** (March-April, 1968)

SINKANKAS, JOHN, 'Correspondence' (October 18, 1976)

SINKANKAS, JOHN, *Gemstones of North America,* **I** (1959)

SINKANKAS, JOHN, *Gemstones of North America,* **II** (1976)

SINKANKAS, JOHN and REID, ARCH. M., 'Colour-composition relationship in spessartine from Amelia, Virginia', *The Journal of Gemmology* (October, 1966)

SKINNER, BRIAN J., 'Physical properties of end-members of the garnet group', *American Mineralogist,* **41** (1956)

SMITH, HERBERT, G. F., *Gemstones,* London (1977)

STERRETT, D. B., 'Garnet', *Mineral Resources* (1908)

STROCK, LESTER W., 'Spessartite from Avondale, Delaware County, Pennsylvania', *The American Mineralogist,* **15,** No. 1, January (1930), 'Notes and News'

TISDALL, F. S. H., 'Spessartites from Madagascar', *The Gemmologist,* **XXXI,** No. 372 (July, 1962)

WEBSTER, ROBERT, *Gems* 4th edition (London, 1983)

Chapter 7
Grossular

Nomenclature and history

Although grossular garnet is a relatively modern term, the yellowish-orange gemstones of the species were known to the ancient world from important sources in Ceylon and India. However, it is not clear whether they were classified under the ancient terms, 'hyacinthus', 'lyncurium', 'carbuncle', or some other category.

While 'hyacinthus' is the name which was eventually assigned to the grossular garnet, it is quite probable that it was not garnet at all in the Roman period. The improbability is suggested by the colour assigned to it by Pliny (41:1): a 'diluted' violet, associated with amethyst, but thought by some to be corundum (Eichholz, 1971, p. 262, note a).

The lyncurium stone, on the other hand, was the right colour for grossular ('amber-coloured'); it was also the right hardness (much harder than amber, difficult to cut); it was also the right usage (seals were cut from it). Nevertheless, it was a stone supposedly produced by an animal, and Pliny scoffed at the idea and repudiated its existence in his day (Pliny 37:14). The animal, the lynx in this instance, produced the stone when its urine mixed with the earth (hence the name 'lyncurium').

Although Pliny strongly denounced its medicinal uses, Marbode prescribed it as a cure for jaundice and diarrhoea in the 11th century (King, 1866). It is also speculated that the colour of the urine matched the gemstones, the colour of the male urine being reddish and fiery, that of the female being pale or even white (Caley and Richards, 1956). Such a colour range is perfectly exhibited by the grossular garnets.

In the writings of Marbode, however, other colours were attributed to 'hyacinth' and the original 'violet' was omitted, or changed to blue. Red was one of its colours, yellow was another and the third colour was blue. The occurrence of these particular colours are thought to represent corundum by a number of modern writers.

However, by Nicol's time (1652), hyacinth was only red, yellow and gold-coloured. Blue was omitted from the colours. Furthermore, sources were suggested in India ('Cananor, Calecut and Cambaia') for the finest material, and also along the river Isera, near Silesia and Bohemia, for poor quality specimens. The transformation of 'hyacinthus' was thus complete. From a variety of amethyst (perhaps rose quartz) in the ancient period, it was reclassified in the Medieval period, possibly to corundum, and finally to grossular garnet or zircon in the 17th century. Also in the same century, the terms 'hyacinth' and 'jacinth' were used interchangeably (Nicols, 1652).

The colours of the hyacinth were described at the beginning of the 17th century revealing colours of red, yellowish-red and amber, as well as near colourless. What is also interesting is that the same author (de Boodt, 1609) gave a unique description of the many inclusions that are so common in the brownish-orange grossular garnet:

> 'these (the amber coloured ones) are of no great value, by reason of the atomes which they do contain, and the multiplicitie of small bodies which are in them, which do hinder their transparencie and disphanity' (de Boodt, 1609, translated, or paraphrased by Nicols, 1652)

Unfortunately zircon also displayed the same colours, so it, too, was included in the category. Consequently, for some time hyacinth or jacinth was the species name applied to both zircon and grossular garnet. But when hardness testing was routinely applied to the stones, it was found that the grossular garnets were inferior in hardness to zircon and a new name was invented: 'hessonite' from the Greek word ἥσσων, meaning less, or inferior. Nevertheless, jewellers still classified zircon and grossular under 'hyacinth', well into the 20th century. Bauer related that jewellers of his day were assigning zircon or grossular to the categories on the basis of colour: the dark colour would be assigned to hyacinth, while the light coloured stones would be designated hessonite (Bauer, 1905, p. 350). Scientifically, however, the gems were categorized as hyacinth for zircon and hessonite for grossular at the end of the 19th century, if not before.

The term, 'grossular' was derived from the gooseberry, which displayed a pale green colour, found to be common among some hessonites. Later, 'grossular' was designated a species name for this family of garnets, a name which it still occupies. The term 'hessonite' is now vaguely assigned to the yellow to orange-red colours of the species, while the original 'hyacinth' is now dropped in favour of the term 'zircon'.

Another name came into use and was applied to a certain colour of the Ceylon grossulars. 'Cinnamon stone' became known as the grossular garnet from Ceylon with a colour that matched that of the 'oil of cinnamon' produced in the same locality (Feuchtwanger, 1867). 'Cinnamon-stone' is still used in some segments of the gem trade, but it is not a recognized variety name, and perhaps it should be deleted from modern gemmological nomenclature.

Grossular colours

In addition to the orange-red to yellow colours of hessonite, the gem exhibits other colours as well. Pink is known to exist in addition to a full complement of colours in the yellow-green to near pure green. Colourless and near colourless samples are also found, although they are reportedly rare in some locations (Muije *et al.,* 1971).

Pure or very nearly pure grossulars do exist in nature, unlike other garnet species. The pure samples are colourless; therefore, grossular is known as an allochromatic gem, which relies on foreign trace elements in varying quantities to impart colour. The colouring agents in grossular seem to be ferric iron, manganese, chromium or vanadium. In a study of the grossular garnets, Manson and Stockton (1982) established a correlation between the Fe_2O_3 and the percentage of the yellow-orange trend. The same element apparently had no influence upon the colour of the green grossulars, however. Furthermore, FeO could not be determined to show a clear-cut relationship in the yellow to orange colours, even though it was present in some of the samples in amounts below 2.0% weight percentage.

In the same study, vanadium was also found to be related to the production of green in the grossulars, but not the yellow to orange colours of the species. Although this fact was already well established by a number of earlier studies on tsavolite, the same positive correlation was noted in chromium, as well, where the chromium was found in amounts up to 0.5% weight percentage. The study by Muije *et al.* (1971) had concluded that chromium was not a factor in tsavolite colouration. However, in the latter study, the sixteen sample stones had no more than 0.2% weight percentage of chromium.

Colour suites from colourless to many shades of green, yellow, orange to orange-red are often seen in the grossular garnets. Many hue positions and tones are the hallmarks of the grossular gemstones.

The green colours also range widely, from very pale (near colourless) greens to intense purish-green. The hue position most commonly found in the green grossulars, however, is a yellowish-green: either strong yellowish-green or on to a near pure green with a minimum of yellow. Very rarely a slightly bluish-green can be found in tsavolite.

The green stones are now classified as either green grossular or tsavolite, depending on the intensity and tone of the green colour, very much like the arbitrary division between green beryl and emerald, or between pink sapphire and ruby. The rich greens of the tsavolite have been likened to the fine colours of high quality emerald, although emerald specialists might reasonably debate the issue. Nevertheless, the high intensities of green in the tsavolite produce a very rich colouration in the gemstone, driving prices up to thousands of dollars per carat for top-quality specimens. Incidentally, in American usage, the term tsavorite is commonly used; but in the UK and Europe tsavolite is the accepted gemmological term.

Grossular chemistry $Ca_3 Al_2 (Si O_4)_3$

Grossular is a calcium-aluminum silicate which is coloured by ferric iron, chromium and/or vanadium, with other elements possibly contributing to some extent. In an attempt to understand grossular composition from natural specimens, seventeen samples were taken from the literature, revealing RI/SG values in addition to

TABLE 7.1. Grossular end-member components and RI/SG properties from grossulars reported in the literature. The high properties reported in this list do not seem to correlate with gem grossulars, which are typically not over 1.755 in RI (see *Table 7.2*)

Samples	1	2	3	4	5	6	7	8	9	10	11	12	13	14	15	16	17
RI	1.728	1.736	1.737	1.737	1.745	1.747	1.747	1.749	1.755	1.756	1.756	1.763	1.769	1.772	1.796	1.801	1.808
SG	3.635	3.506	3.506	3.600	3.611	3.650	3.599	3.512	3.629	3.567	3.620	3.630	3.688	3.770	3.676	3.710	–
End-member components in percentages from chemical analyses:																	
Grossular	69.00	94.66	96.00	96.60	87.88	85.70	79.20	80.85	87.29	85.21	83.96	79.19	71.00	59.10	51.20	54.20	60.00
Andradite	26.90	0.57	2.30	–	2.03	8.00	9.50	7.88	12.71	12.88	10.80	20.51	16.70	22.40	7.60	42.90	30.00
Uvarovite	–	–	0.40	–	–	–	–	–	–	–	–	–	–	–	35.90	–	–
Spessartite	1.20	2.49	–	0.32	0.71	2.80	0.10	0.14	–	0.32	0.89	0.30	1.40	5.00	–	2.20	–
Almandine	1.80	0.99	0.10	2.06	8.07	3.30	5.80	3.82	–	0.52	3.63	–	7.40	13.60	–	0.50	5.00
Pyrope	–	1.29	–	1.02	1.31	–	5.00	7.32	–	1.07	0.72	–	2.90	–	0.80	–	5.00
Other	1.10	–	–	–	–	0.20	0.40	–	–	–	–	–	0.60	–	0.90	0.20	–

1. Hyacinth-red grossular, Monte Rosso di Verra, Italy; Deer, Howie and Zussman, 1962, recalculated by Rickwood (1968)
2. Grossular from Xalostoc, Mexico; Ford (1915)
3. Grossular from Georgetown, California; Fleischer (1937) recalculated by Rickwood (1968)
4. Grossular from the Black Lake area, Quebec; Fleischer (1937)
5. Grossular from Maigelstal, Switzerland; Fleischer (1937)
6. Gem grossular from Umba, Tanzania; Jobbins, *et al.* (1978) (reddish-brown colour)
7. Reddish-yellow grossular, Val d'Aosta, Italy (Sanero, 1935); Deer, Howie and Zussman (1962) recalculated by Rickwood (1968)
8. Grossular from Passa del Termine, Adamello, Italy; Fleischer (1937)
9. Grossular from Italian Mt., Gunnison County, Colorado; Fleischer (1937)
10. Grossular from Sierra Tlayacac, Mexico; Fleischer (1937)
11. Grossular from West Redding, Connecticut; Fleischer (1937)
12. Grossular from Ala, Piedmont; Ford (1915)
13. Brownish-red grossular from Salem district, Madras, India; Deer, Howie and Zussman (1962) recalculated by Rickwood (1968)
14. Red-brown grossular, Milendella, South Australia; Deer, Howie and Zussman (1962) recalculated by Rickwood (1968)
15. Chromian grossular from pyrrhotite quartzite, Luikenlahti, Finland; Deer, Howie and Zussman, 1962, recalculated by Rickwood (1968)
16. Grossular, South Africa; Deer, Howie and Zussman (1962) recalculated by Rickwood (1968)
17. Grossular from Bushveld, South Africa; Wright (1938)

end-member compositions based on chemical analyses (*Table 7.1*). Although these stones may or may not represent gem quality specimens, nevertheless they do represent natural grossulars. These stones range in composition from 51.2% grossular to a high of 96.60% grossular. Although they are reported to be in solid solution (Phillips and Griffen, 1981), there is little miscibility seen between uvarovite and grossular in the samples of *Table 7.1*, with the one rare exception is example No. 15. It is a uvarovite-grossular from Finland with a very high percentage of uvarovite (35.9%).

The most common end-member component seen in *Table 7.1*, is the andradite molecule. It exists in all of the samples except No. 4, which is near pure grossular (96.60%). The miscibility is very much in evidence between andradite and grossular, for the percentages range from 0.57% to 42.9% andradite. Such stones are called 'grandite'.

The strong miscibility between grossular and andradite also influences the RI values (*Figure 7.1*). Moreover, the high solubility of the two end-members even produces an intermediate category, similar to the pyrope-almandine classification. But, judging from the literature, they are much more rare than the pyralspite intermediates. Wright (1938) published one grandite (RI 1.804) with 50% grossular, 40% andradite, 5% pyrope and 5% almandine; another is cited with only 45% grossular, 40% andradite, 5% pyrope and 10% almandine with an RI of 1.809.

There may also be a strong miscibility between the almandine end-member and the grossular. Although the end-member percentages of the almandine in *Table 7.1*

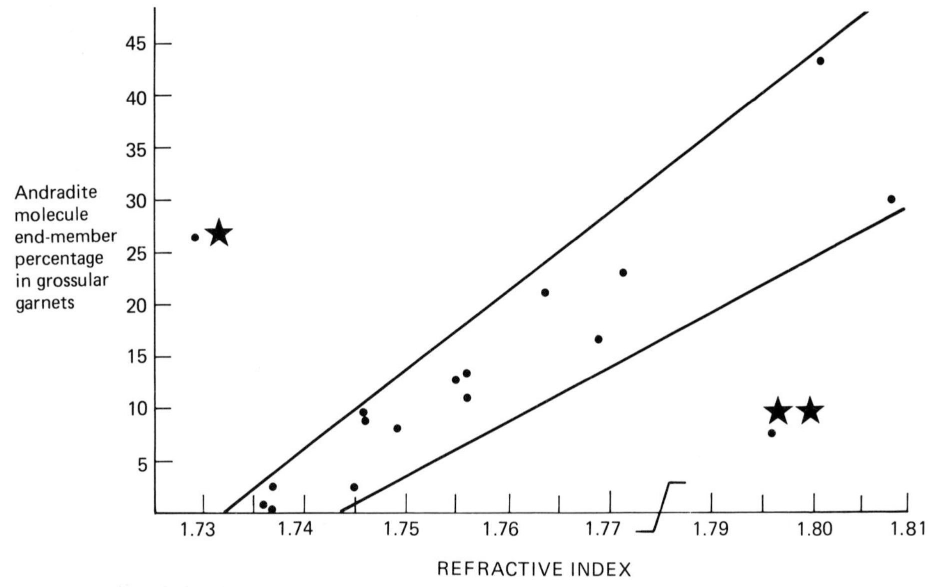

Figure 7.1 The relationship of the refractive index to the end-member component of andradite, from data in *Table 7.1*

* This stone represents an anomaly. With over 25% andradite, one would expect a much higher RI. We cannot explain it, unless there was an error in the calculation, or a misprint

** This stone exhibits a low andradite component, but the high RI is due to the high percentage of uvarovite (35.9%)

are relatively low (to 13.6%) there are others reported in the literature with much higher concentrations. For example, Wright (1938) reported an almandine-grossular with 59% grossular and 39.5% almandine, and another with 72.2% grossular and 20.1% almandine. So there is some miscibility in evidence, although the rarity of examples is most obvious.

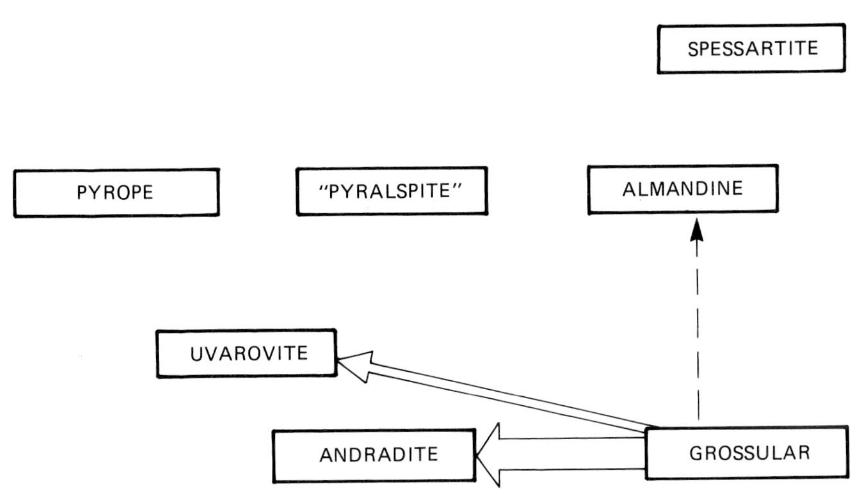

Figure 7.2 The miscibility relationship of grossular with other garnet end-members. Grossular and andradite are in solid solution with each other; uvarovite and grossular are also in solid solution, even though few examples in *Table 7.1* reveal the relationship. Enough almandine-grossular was found in the literature to reveal a high correlation of solubility between them. Pyrope and spessartite provide only limited miscibility with grossular, although replacement of Mn can occasionally modify grossular colouration

A much smaller miscibility relationship exists between grossular and either pyrope or spessartite. The percentages in *Table 7.1* for both pyrope and spessartite reveal a high of only 5% in the samples. Wright (1938) exhibited just three stones with higher pyrope percentages, one at 6.9%, another at 6.7% and a third at 10%. The highest spessartite percentage in Wright's list is just 3.5%. The miscibility relationships between grossular and the other end-member components are illustrated in *Figure 7.2*.

Optical and physical properties

The optical and specific gravity values of grossular seem to vary considerably. Webster (1983) limited the RI of gem grossulars to a small range of 1.742 to 1.748. His SG was listed at 3.65. Smith's RI and SG values for grossular were reported at 1.741–1.748 and 3.60 to 3.80, respectively (Smith, 1972, p. 332). In Manson and Stockton's study (1982), their study stones, from unknown sources, ranged from 1.733 to 1.760 with SG properties from 3.59 to 3.66.

In the stones collected from the literature (*Table 7.1*), a much broader range of properties can be seen. Examples are reported from 1.728 to 1.808 for RI values. The stone with the lowest value was taken from Deer, Howie and Zussman (1962)

as their sample No. 5. The RI was quite exact, for it was measured to the third place (1.7284), although it seemed unusually low, particularly for a 'hyacinth-red' coloured sample. However, we have seen colourless gem grossulars with a refractive index of 1.730–1.734. The specimen at 1.730 was very nearly completely colourless while those with very slight tints ranged higher in RI. Pure synthetic grossular is reported to be 1.734, with an SG of 3.594 (Skinner, 1956).

The stones of *Table 7.1* can be seen to increase in RI as the andradite component increases (*Figure 7.1*). There are two exceptions in the chart, however. One is an anomaly (No. 1, described above), and the other is attributable to a high uvarovite molecule (No. 15). The high values of the RI seem to indicate a problem, for gem grossulars are traditionally reported to be lower than 1.750 (*Table 7.2*). A full 50% of the stones in *Table 7.1* are above 1.750. It is either that known gem occurrences produce little or no high-property 'grandites' or that the samples quoted are extremely rare specimens. Another possibility exists: Manson and Stockton's study revealed a range of RI values to 1.760; it is quite conceivable that future RI studies will find other samples of gem material to widen the range still further.

TABLE 7.2. Comparison of optical and density values reported by traditional sources and tests conducted on recent gem specimens from known and unknown sources

Source, date	*Samples Species/variety*	*RI*	*SG*
Arbunies, *et al.*, 1975	(13) Light green gross. Italy	1.732–1.738	3.58–3.63
Arbunies, *et al.*, 1975	(40) Hessonite, Canada	1.738–1.748	3.43–3.78
Muije, *et al.*, 1979	(3) Colorless grossular, Tanzania	1.731–1.732	3.61–3.62
Jobbins, *et al.*	(1) Reddish-brown grossular, Tanz.	1.747	3.650
Manson and Stockton, 1982	(16) Hessonite, source unknown	1.733–1.760	3.59–3.66
Manson and Stockton, 1982	(105) Grossular, source unknown	1.731–1.754	3.57–3.67
Gübelin and Weibel, 1975	(?) Tsavolite, Kenya	1.739–1.744	3.57–3.65
Frankel, 1959	(5) Hydrogrossular, So. Africa	1.708–1.731	3.401–3.523
Webster, 1983	Traditional values	1.742–1.748	3.65
Smith, 1972	Traditional values	1.741–1.748	3.60–3.80

Consequently, it should be understood that the high range of RI in grossular garnet is not fixed, but it may be seen to follow the minerals which do exist in the higher RI ranges. High property 'grandites' can still be separated from the other garnets, however, by means of SG analysis, absorption spectrum and even colour. The stones in the very high RI ranges of *Table 7.1*, therefore, should be viewed as rarities at the moment, which may or may not have their counterparts in gem samples from known sources.

The specific gravity range of the stones in *Table 7.1* is from 3.506 to 3.77, representing slightly lower properties than those quoted by traditional sources (3.80, Smith, *Table 7.2*).

Ugrandites are often found showing some degree of birefringence. Studies have suggested that the cause for such an unusual feature in isotropic minerals is to be found because of residual strain in the structure, or perhaps ordering of octahedral cations, Fe^{3+} (ferric iron) and Al, thereby reducing the symmetry (Akizuki, 1984).

In the latter study the author analyzed samples of such grandites from Eden Hill, Vermont and Fujikaramari, Japan. The Vermont stones were described as transparent grossulars of 'cinnamon' colour up to 5 mm in diameter. There was no mention of their optical or density properties or whether these stones were of gem quality (beyond the above description). They were, on the average 85.7% grossular and 14.3% andradite. The surface features were studied and were found to correlate well with the internal texture, thereby occurring during growth of the crystal. The grossulars from Japan were brownish-red and translucent. They were 79.7% grossular and 20.3% andradite and revealed growth sectors appearing as fine lamellae under the microscope.

The birefringence most often found in ugrandites is rarely over 0.005 (Phillips and Griffen, 1981). However, one very high sample was reported to have a high birefringence of 0.012 (Winchell and Winchell, 1968, p. 490).

The dispersion for grossular garnet is reported at 0.028 (Smith, 1972), which is only moderate in strength. Even in the colourless stones, therefore, the spectral fire admired so much in diamond (0.044) will not be significant in grossular.

The hardness of grossular also varies, but it is generally considered to be slightly above 7 on Mohs' scale. However, the study by Arbunies *et al.* (1975) indicated a range from 6.66 to 6.98 for the light green grossular gems from Italy, and 6.46 to 6.90 for the Canadian hessonite gemstones.

Grossular garnets generally do not exhibit characteristic spectra, but some may show a very faint almandine spectrum due to the FeO from the almandine molecule when it is present as a trace element. Several stones of the Manson and Stockton study (1982) revealed a faint band at 434 nm which was attributed to Fe^{3+}; another stone of light orange colour exhibited weak bands at 418.5, 489.5, 503.5 and 529.5 nm, all possibly related to inclusions, but most of the yellow to orange stones revealed no sharp absorption bands.

Grossular inclusions

The internal features of grossular gemstones are very pronounced and are characteristic for the gemstone. The heat-wave effect, or treacle is uniformly associated with hessonite. Its swirly appearance is very distinctive and is often seen. Liquid filled inclusions and negative crystals are also common, although not necessarily characteristic to grossular, either by their presence or pattern. Crystals are also found, commonly distributed throughout the whole stone. Such crystals are reportedly diopside and zircon (Gübelin, 1953).

Two-phase inclusions (liquid and gas) are also common inclusions in grossular. Treacle may or may not be seen in colourless grossular, but 'short needles and angular inclusions' as well as fingerprint-types are reported (Muije *et al.*, 1971). In a study of hessonites from Harts Range, Northern Territory, Australia, no treacle was found, but pyrite crystals with planar stress cracks were observed (Bracewell and Brown, 1983). There was also an absence of masses of crystal inclusions common to other grossulars. Sometimes the diopside inclusions take on strange shapes. Charles Fryer of the GIA photographed an unusual specimen (source not noted) revealing 'ladder-like' forms in a hessonite (*Gems and Gemology*, Fall, 1974).

Sources for hessonite and colourless grossular

An ancient source for grossular garnets is India. Fermor (1938) reported that grossular was not uncommonly found in the Central Provinces. Iyer (1961) mentioned grossular from Madras in the Coimatore district, ranging in colour from yellowish-brown to light yellow and yellowish white in good to fine specimens. Orissa produced 'cinnamon brown to light brown coloured' grandites, found in kodurite near Boirani in Genjam district; another source from a pegmatite at Daolathgarh in Rajasthan is also cited (Iyer, *op. cit.*).

A major source of hessonite has also been Ceylon. In addition to the India sources, above, the Ceylon source provided the Greeks and the Romans with their 'lyncurium' stones. Feuchtwanger (1867) reported that in his day hessonite was so much associated with Ceylon that it was called by French jewellers, 'hyacinth de Ceylon'. Even Bauer (1905) and Farrington (1903) describe Ceylon as the only major source for gem hessonites. Sri Lanka (Ceylon) still remains today as one of the major sources for gem hessonites in the world.

Miers (1902), however, failed to mention the important Ceylon sources, but listed numerous European sources for the mineral, including Ala (Turin) in brilliant 'hyacinthe-red' colours, Vesuvius in lustrous yellow or brown crystals, Achtaragda River in Siberia, and others. Farrington (1903) claimed that those in Italy (Ala, etc.) were too small for gemstones, but that they made pretty mineral specimens. Bauer confirmed that the Ala hessonites were beautiful as specimens, since they occurred as druses and were associated with dark green chlorite and pale green diopside, but that they were too small for cutting (Bauer, 1905, pp. 351–352).

However, in recent years, a light green grossular has been found in Val de Fassa, Italy which can cut gems from 4 mm to 5 mm in size (Arbunies *et al.*, 1975). Also, some hessonites are found in Brazil, in the Minas Gerais mining district (Webster, 1983). Another source, perhaps not of gem material, however, is a brownish-green grossular found along the Vilui River in Siberia (Webster, *op. cit.*).

Along with the important finds of tsavolite in Tanzania-Kenya border areas in the late 1960s, an important source was simultaneously discovered in the colourless, yellow and orange grossulars as well (Muije *et al.*, 1971). East Africa is still a major source of colourless grossulars and hessonites.

Mexico has also been found to produce very fine orange hessonites of gem quality. Recent workings of these mines have resulted in a profusion of hessonites in the gem marketplace, although usually confined to five or ten dealers. Recent conversations with the dealers of these gems, however, reveal that there may be some scarcity in the near future, as the mines are no longer being worked to the same extent as they have been in previous years.

In studying large samples of this material from the mine owner in 1981, most stones of gem quality were found to be under 10 carats in size. Very few stones were found that would cut 4–5 carat gems. Moreover, colours ranged from light 'cinnamon' to a delightful orange, somewhat similar to spessartites of gem quality. However, there was always the slight brownish tint pervading the body colours, characteristic of the hessonites.

Grossular garnet is also found in North America. A very famous deposit is the Jeffrey Mine in Quebec, Canada. It has been in operation since the 1880s, producing largely chrysolite asbestos, although there are about 66 minerals which have been discovered in the site, including the hessonite garnets. The garnets occur in many colours, from colourless through to white, pink, orange and green (Wight and Grice, 1982). The green grossulars from this site are most rare, with only one gemstone reported in the National Museum's collection (weighing 0.25 carat). Six faceted gems weighing from 1.64 to 7.08 carats were studied in 1981: these ranged in colour from almost colourless to slightly yellow and exhibited a single RI for all six stones, 1.733 (Wight and Grice, 1981).

The pink colours have not been found in specimens large enough, or in enough quantities for use as gemstones; their unusual colour was attributed to trace amounts of Mn (0.5% weight percentage; Wight and Grice, 1981). Small gems in a slightly pale cinnamon colour (1–3 carats) have been marketed and continue to be seen occasionally in the gem marketplace.

Another asbestos mine in Vermont has recently been reported as a gem producer of hessonites. The mines are in Lamoille County, north of Eden Hills; although these stones are of a 'beautiful' colour, they are only large enough to cut very small stones (Sinkankas, 1976, p. 189). The Pala District of San Diego County, California has also produced some grossular garnets from time to time. Kunz (1890) mentioned many US localities of gem hessonite, including Phippsburgh, Maine; Warren, New Hampshire; Avondale, Pennsylvania; Bakersville, N.C. and others (p. 79). Green grossular was also reported by Kunz, occurring in rich, dark green stones of about ½ inch in diameter from the Tilly Foster Mine in Brewster, New York. In another site in Sonoma County, California a green grossular was reported, as well as the Good Hope Mine in California; however, these are isolated references and probably isolated occurrences as well, even if the stones are properly classified (Kunz, *op. cit.*, p. 79).

Tsavolite garnet

The first 'emerald' green grossular garnets were found in Tanzania in the late 1960s (Muije *et al.*, 1971). Bank *et al.* (1970) studied them and declared that their unique colour was due to chromium occurring as a trace mineral. The find was in the Lelatema Mountains of Tanzania where the colourless, yellow and brown grossulars were found in addition to the green grossulars.

However, two other parties also were involved in prospecting in Kenya for similar material. Geologist Campbell R. Bridges located deposits of green grossular garnet in 1971, in an area just south of the Teita Hills in Kenya. The deposit also contained a certain amount of green tourmaline, blue zoisite (later named tanzanite), and the green grossular (Bridges, 1974). The colours were described as ranging from colourless, 'through pale lime green to a full rich grass or emerald green' (Bridges, *op. cit.*). The occurrence was noted to be in a graphite gneiss.

Another discovery of the green garnets was made in 1973 by a prospector hired by Kimani and Morgan, who led the expedition. It proved to be a prolific deposit and was named the Lualenyi Mine (Gübelin and Weibel, 1975). Analysed samples revealed an RI range of 1.739 to 1.744, with the paler colours in the lower range; an increase of vanadium revealed a higher refractive index. The SG indicated a range from 3.57 to 3.65. The chemical analysis seemed to prove that the green colour was due solely to the vanadium occurring as a trace element in the specimens, rather than chromium, which was attributed to the Tanzanian green grossulars.

Absorption bands were found, corresponding to the vanadium content, in two broad bands centring at 430 nm and 610 nm. Inclusions in the vanadium-rich grossulars revealed veil-like feathers made up of liquid inclusions, negative crystals and solid crystals. One crystal inclusion was determined to have a composite structure which included quartz, calcite, chrysotile or enstatite and alabandite. Another unusual discovery was byssolite-like fibres arranged in bundles or in isolated occurrences. The fibres, however, did not follow that 'horse-tail' pattern common to demantoid garnets. The authors of this study (Gübelin and Weibel, 1975) also suggested that the name for the material be 'vanadium grossular garnet'.

The name 'tsavorite' (later changed to 'tsavolite') was introduced by Tiffany & Co. in September, 1974. The name was taken from the Tsavo National Park, very close to the original Kenya deposits and was tied to a large publicity campaign by Tiffany's to promote the gemstone.

The sizes of the tsavolites have not been large. A one carat cut specimen is considered somewhat rare; one to three carats are not uncommon, but their prices are high, reflecting the increased rarity; over three carats are very rare and command prices that can be found in the thousands of dollars per carat. The largest tsavolite we know of is a 20 carat fine gem owned by a dealer specializing in the African gems.

As gemstones, tsavolites are considered to be one of the most attractive of the garnets. Furthermore, they have the hardness (over 7 on Mohs scale), durability (they are not brittle), high optical properties to produce a brilliant gemstone, and relative clarity (compared to emeralds) necessary to establish them as classic gems. Furthermore, the market for the tsavolite today is somewhat controlled in Nairobi; the supply is not abundant and the dealers are controlling it so that there is not a large supply dropped on a slack market. This control of the market is rare in coloured gemstones and usually only functions when there is one source of the material; however, it is much like the DeBeers control over diamond supply and demand, created to benefit miner, dealer and consumer from erratic pricing fluctuations.

Tsavolite colours in gemstones seem to exhibit high preferences when the hue position is found close to pure green; as the colour drifts into the yellow side of green the appeal drops correspondingly. Very light tones and very dark tones are also not as highly valued as the stones situated in the mid tonal range. Bright, vivid intensities of near purish green in the medium tones are features most viewers admire and appreciate in tsavolite.

Hydrogrossular

Another form of grossular garnet that has been cut into gems is the massive hydrogrossular. A report of the material, found in the African Bushveld and called 'Transvaal jade' because of its jade-like appearance, was made by Hall (1924). It was described as a translucent to opaque compact rock with a waxy lustre and somewhat conchoidal, or splintery fracture, generally revealing inclusions of black specks. The black specks were magnetite, or in some cases, chromite (Webster, 1983). Hall classified the colours into several types – a green colour varying between light and deep green with some yellow overtones, a pale grey, a very delicate 'mother-of-pearl' colour ranging from pure pinkish to faint bluish phases, and occasionally dull greys, pale cream colours and pale, or dirty, whitish material (Hall, 1924). He described the hardness between 7 and 8, with the green types in the higher hardness category. The SG range was reported to be from 3.33 to 3.52. Samples tested showed anomalous double refraction and chemical analysis proved that the material was not jade at all, but grossular garnet with some H_2O present. The cause of the colour was due to chromium (green samples) or manganese (pink specimens).

Frankel (1959) also studied samples of the hydrogrossular from the same source. An RI range of 1.708 to 1.731 and SG range of 3.401 to 3.523 was seen to be correlated with the water content. The lower RI/SG properties revealed the highest water percentages. Furthermore, the pink colours tended to be found in the lower RI and SG values, whereas the green colours reflected higher properties (Webster, 1983).

Webster also noticed a characteristic X-ray luminescence, quite unlike jadeite, nephrite or any other jade simulant, with a strong orange-yellow colour emitted. Massive hydrogrossular also forms a solid solution with the mineral idocrase, and variations in chemical composition can and do occur between the two end-members. Other sources for this garnet are New Zealand, Utah, Pakistan and Alaska; but the finest greens in gem quality specimens have come from the South Africa source.

In the finest gem specimens, the translucency is very high, approaching transparency, and the green colour is exceptionally intense pure green, resembling the finest jade. The magnetite or chromite inclusions are missing from the finest examples, but are quite common in lower qualities. The lustre is also very high in the finished gems, adding to their beauty.

Unusual phenomenon grossular

Garnets are full of surprises. An unusual stone was found in Nevada, about a mile east of the Adelaide mining district in north-central Nevada. The garnet was described by Ingerson and Barksdale (1943), from whose account the following information is taken. Field tests were made to determine the identity, and subsequent laboratory tests showed the material to be grossular with considerable andradite, revealing an RI of 1.81 and SG of 3.50 to 3.64. Colours ranged from 'honey-coloured' to dark brown. The material was mostly massive in form, but many crystals were also found. The most striking feature of these garnets was a brilliant play of colours seen in reflected light on the striated compound faces of most of the lighter crystals. The spectral colours were seen in bands parallel to the surface striations, but the phenomenon was not due to surface film. Photos were included which illustrated the iridescence on a crystal face and also in thin section.

Under the microscope the stones were found to be slightly birefringent and also zoned. Superimposed upon the coarse zoning was a fine lamellae in a herringbone pattern which concentrated in the core of the crystal, rather than the outer surface. In addition to the fine lamellae, the cores exhibited a complex mottling, not unlike some microcline twinning.

Experiments with heat reduced the birefringence considerably, to a very weak state, but the iridescence remained strong. Even when the temperature was brought up to the very point of melting the sample, the iridescence still remained as strong as ever. The cause of the iridescent colouration was suggested to be from the very fine polysynthetic twinning and the intensity of the iridescent colours was dependent upon the lamellae thickness.

In a recent study of the stone (Hirai *et al.*, 1982), the authors found that the cause of the iridescence was due to compositional variation within the lamellae. Iron-rich lamellae were composed of $An_{92}Gr_7Sp_1$, while the iron-poor lamellae were composed of $An_{88}Gr_{10}Sp_2$. Whatever the cause of the phenomenon, it certainly highlights an unusual garnet.

Summary of grossular properties

Chemistry:	$Ca_3\,Al_2\,(Si\,O_4)_3$
Colouring agents:	Green: vanadium and/or chromium; other colours: ferric iron, Mn
Refractive index:	1.730–1.760
Specific gravity:	3.40–3.78
Absorption spectra:	No characteristic spectrum
Hardness:	7
Colours:	Many colours, including pink, orange-red, brownish-red, orange, yellow, green and colourless
Dispersion:	0.028

Bibliography

AKIZUKI, MIZUHIKO, 'Origin of optical variations in grossular-andradite garnet', *American Mineralogist,* **69** (1984)

ANDERSON, B. W., 'Properties and Classification of individual garnets', *The Journal of Gemmology,* **VII,** No. 1 (January, 1959)

ANDERSON, B. W., 'Transparent green grossular – a new gem variety; together with observations on translucent grossular and idocrase', *The Journal of Gemmology,* **10,** No. 4 (October, 1966)

ARBUNIÉS, ANDREU M., BOSCH-FIGUEROA, J. M., FONT-ALTABA, M. and TRAVERÍA-CROS, A., 'Physical and optical properties of garnets of gem quality', *Fortschr. Miner.,* **52,** Stuttgart (December, 1975)

BANK, H., BERDESINSKI, W. and OTTEMAN, J., 'Durchsichtiger Smaragdgrüner grossular aus Tansania', *Z. Dt. Gemmol. Ges.,* **19** (1970)

BANK, H., 'Über grossular unt hydrogrossular', *Z. Dt. Gemmol. Ges.,* **31** (1982)

BAUER, MAX, *Precious Stones,* Vermont and Tokyo (1970 from the 1905 edition, Trans. by L. J. Spencer

BRACEWELL, H. and BROWN, G., 'Harts Range Hessionite (sic.)', *The Australian Gemmologist* (February, 1983)

BRIDGES, CAMPBELL R., 'Green grossularite garnets ("Tsavorites") in East Africa', *Gems and Gemology* (Summer, 1974)

CALEY, EARLE and RICHARDS, JOHN F. C., *Theophrastus on Stones,* Columbus, Ohio (1956)

DE BOODT, BOETIUS, *Gemmarum et Lapidum* (1609)

DEER, N. A., HOWIE, R. A. and ZUSSMAN, J., Rock-Forming Minerals, *Ortho and Ring Silicates,* **I,** London (1962)

EICHHOLTZ, D. E., *Pliny, Natural History,* Vol. X, Cambridge, Mass., Loeb Classical Library (1971)

FARRINGTON, OLIVER CUMMINGS, *Gems and Gem Minerals,* Chicago (1903)

FERMOR, SIR LEWIS LEIGH, *Garnets and Their Role in Nature,* Calcutta (1938)

FEUCHTWANGER, L., *A Popular Treatise on Gems,* New York (1867)

FLEISCHER, MICHAEL, 'The relation between chemical composition and physical properties in the garnet group', *The American Mineralogist,* **22,** 6 (June 1937)

FORD, W. E., 'A study of the relations existing between the Chemical, Optical, and other physical properties of the members of the garnet group', *American Journal of Science,* Fourth Series, Vol. XL, No. 235 (July, 1915)

FRANKEL, J. J., 'Uvarovite garnet and South African jade (hydro grossular) from the Bushveld Complex, Transvaal', *The American Mineralogist,* **41** (May-June, 1959)

FRYER, CHARLES, 'Unusual inclusions', *Gems and Gemology,* Fall (1974)

GRICE, J. D. and WILLIAMS, R., 'The Jeffrey Mine Asbestos, Quebec', *The Mineralogical Record* (March-April, 1979)

GUBELIN, EDWARD J., *Inclusions as a means of Gemstone Identification,* Santa Monica, California (1953)

GUBELIN, E. J. and WEIBEL, M., 'Green vanadium grossular garnet from Lualenyi, near Voi, Kenya', *Lapidary Journal* (May, 1975)

HALL, A. L., 'On "Jade" (massive garnet) from the Bushveld in the Western Transvaal', *Transactions of the Geological Society of South Africa* (1924)

HIRAI, HISAKO, SUENO, SHIGEHO and NAKAZAWA, HIROMOTO, 'A lamellar texture with chemical contrast in grandite garnet from Nevada', *American Mineralogist,* **67** (1982)

INGERSON, EARL and BARKSDALE, JULIAN D., 'Iridescent garnet from the Adelaide Mining District, Nevada', *American Mineralogist,* **28** (1943)

IYER, L. A. N., 'Indian precious stones', Bulletin No. 18, *Geological Survey of India* (1961)

JOBBINS, E. A., SAUL, J. M., STATHAM, PATRICIA M. and YOUNG, B. R., 'Studies of a gem garnet suite from the Umba River, Tanzania', *The Journal of Gemmology,* **XVI,** 3 (1978)

KUNZ, GEORGE FREDERICK, *Gems and Precious Stones of North America,* New York (1890)

KUNZ, GEORGE FREDERICK, 'Garnet', *Mineral Resources* (1903)

KUNZ, GEORGE FREDERICK, 'Garnet', *Mineral Resources* (1904)

MANSON, D. VINCENT and STOCKTON, CAROL M., 'Gem-quality grossular garnets', *Gems and Gemology* (Winter, 1982)

MANSON, D. VINCENT and STOCKTON, CAROL M., 'Gem garnets: the orange to red-orange Color range', *International Symposium Proceedings* (1982), ed. by Dianne M. Eash

MIERS, HENRY A., *Mineralogy* (1902)

MITCHELL, R. KEITH, 'African grossular garnets; blue topaz; Cobalt Spinel; and grandidierite', *Journal of Gemmology,* **XV,** 7 (1977)

MOHS, FREDERICK, *Treatise on Mineralogy,* **2** (1825)

MUIJE, PIETER, MUIJE, CORNELIUS S. and MUIJE, LILLIAN E., 'Colorless and green grossularite from Tanzania', *Gems and Gemology* (Summer, 1971)

NICOLS, THOMAS, *A Lapidary or the History of Pretious Stones: with Cautions for the undeceiving of all those that deal with Pretious Stones,* Cambridge (1652)

PHILLIPS, WM. REVELL and GRIFFEN, DANA T., *Optical Mineralogy, The Nonopaque Minerals,,* San Francisco (1981)

POHL, W. and NIEDERMAYR, G., 'Geology of the Mwatate Quadrangle (Sheet 195/2) and the vanadium grossularite deposits of the area', *Austromineral Ges. m.b. H.* C/mines and Geological Dept., Nairobi, Kenya (1978)

SCHMETZER, K. and BANK, H., 'Gelbgrüner Grossular aus Ostafrika', *Z. Dt. Gemmol. Ges.,* **31** (1982)

SINKANKAS, JOHN, *Gemstones of North America,* **II** (1976)

SKINNER, BRIAN J., 'Physical properties of end-members of the garnet group', *The American Mineralogist,* **41** (1956)

SMITH, HERBERT G. F., *Gemstones,* London (1977)

SWITZER, GEORGE S., 'Composition of green garnet from Tanzania and Kenya', *Gems and Gemology* (Summer, 1974)

WEBSTER, ROBERT, *Gems,* 4th Edition (1983)

WIGHT, WILLOW and GRICE, J. D., 'Colourless grossular and green Vesuvianite gems from the Jeffrey Mine, Asbestor Quebec', *Canadian Gemmologist* (1981)

WIGHT, WILLOW and GRICE, J. D., 'Grossular garnet from the Jeffrey Mine, Asbestos, Quebec Canada', *Journal of Gemmology,* **XVIII,** 2 (1982)

WINCHELL, ALEXANDER N. and HORACE WINCHELL, *Elements of Optical Mineralogy,* New Delhi (1968)

WRIGHT, W. I., 'The composition and occurrence of garnets', *American Mineralogist,* **23** (1938)

Chapter 8

Andradite and the rare garnets

Andradite is a species of garnet named after M. d'Andrada, a Portuguese mineralogist who described and named 'allochroite' in 1800. Allochroite was a brown or reddish-brown coloured garnet mixed 'with minerals foreign to the species' (Mohs, 1825), and was later assigned to the andradite species as a manganese-iron-calcium-aluminum silicate (Dana, 1932, p. 594). Under the modern species of andradite, many varieties are found, including the titanium-rich melanites and schorlomites as well as the gem varieties of green demantoid and yellow topazolite.

Schorlomite, the black variety, has been used for mourning jewellery. Kunz (1890) reported a substantial amount of the material from Magnet Cove, Arkansas: 'On cutting it yields a dead black stone, having the lustre not quite as metallic as that of rutile, but rather between it and black onyx'. Stones in the deposit were thought to yield sizes up to 20 carats, cut. Separation of schorlomite or melanite from other black stones is not difficult. The high RI of the material 1.872–1.935, and its characteristic specific gravity (3.78–3.90) will serve to separate these black garnets from other black stones rather easily. Moreover, the titanium-rich black garnets are rarities in the modern gem market since 'black onyx' is the most prevalent black stone used in jewellery applications in modern times.

Other varieties of andradite, however, are quite significant today for their rarity, their beauty and their unusual gemmological features. One such stone is demantoid.

Demantoid garnet

The name 'demantoid' is derived from the Dutch term 'demant' meaning diamond, in allusion to its high dispersive powers (Gill, 1978). It was discovered in 1868 in the Urals of Russia and produced considerable interest. It was sold as 'olivene' or 'Uralian emerald' and the majority of the gems were consumed within Russia as late as 1896, when Bauer wrote his treatise. However, Bauer did not think the stone would achieve classic status due to the small sizes in the Ural deposit and its lack of hardness.

Demantoid is another major garnet that is green in colour, in addition to the modern tsavolite. It was the first green garnet to be discovered and it enjoyed a chequered history vying with the classic emerald and the common peridot, with which stones it was commonly associated. Unlike tsavolite, however, there were features of the stone that would mitigate against its popularity as a classic gemstone.

Bauer described two; the small sizes and the lack of hardness. However, a third factor was the rarity of the material. In order for any new gem material to be considered a 'classic' gemstone, enough quantities of gem material must be found to circulate around the world, thereby acquainting jewellers and consumers with its unique features. When gemstones are unique to an area and are found in limited supply, they usually attract little worldwide attention, except to collectors who search out fine examples. Demantoid garnet has now been relegated to a collector stone, even if at one time there may have been sufficient supplies for a worldwide consumer market – a doubtful event.

However, the collector gem market is a thriving market in its own right. Rare stones are sought out and added to collections around the world. Prices paid for such stones might be very low, or into the thousands of dollars per carat, depending on 'what the market will bear'. The stones circulating in this special marketplace have been cut into gemstones, regardless of their suitability as jewellery stones. Indeed, most of the collector stones will never be mounted into jewellery. They are valued for their special rarity, or for other features they may exhibit. Their beauty may or may not lie in their colour; some are colourless, and others are quite unattractive, aesthetically.

For these reasons, demantoid is still sought after and remains a viable gemstone, even if the colour is less than ideal. Demantoid is a most unusual garnet in many respects; prices asked for this unique gemstone have been reported from $750 to $1200 per carat in the moderately priced atmosphere of the 1977 gem marketplace (Gill, 1978). Reasons for this demand will be better understood when its unique properties are explored.

Demantoid chemistry: $Ca_3Fe_2(SiO_4)_3$

In a recent study of twenty-one andradites of gem quality, Manson and Stockton (1983) found the samples exhibited a high degree of purity (97.02% to 99.67% pure andradite). This, then, is another example of a garnet which can be found in nature in a near pure state.

Demantoid is a calcium-iron silicate with trace elements in isomorphous replacement (typically, titanium, chromium, vanadium and aluminum). The colouration is due to the iron content, which occurs in the ferric (Fe_2O_3) and not ferrous (FeO) state, when the yellow dominates over the green. When the green dominates over the yellow to produce the 'grass green' colour, chromium is the colouration agent, even though it can occur in small percentages.

The only garnet end-members found to be relevant in the Manson and Stockton study were schorlomite and uvarovite, representing the titanium and chromium contents, respectively. Titanium may also influence the colour of demantoid, although the microprobe could not give enough accuracy in such low-level trace elements; manganese is also thought to influence colour when it is present, but it was found in one stone only in the parcel (Manson and Stockton, 1983).

Unfortunately, the Manson and Stockton study, extensive as it was, failed to include any fine green samples, which might have provided much more data on the chemistry and the colouration features. A later report by them of three fine green samples did present an interesting paradox. The Cr_2O_3 content was reportedly only 0.14% weight percentage, which was no higher than their yellow-green samples reported earlier (Manson and Stockton, 1984). It can be seen, however, in a study of a recent source of andradite from San Benito County, California, (Payne, 1981), that the chromium element (Cr_2O_3) was found in a much higher weight percentage than those reported by Manson and Stockton (0.97% compared to 0.14%). The sample analysed was a demantoid garnet of a green colour which apparently dominated over the yellow (approximately the 'emerald green' colour).

Optical and physical properties

Demantoid is quite remarkable for its capacity to return light in brilliance. Brilliance in gemstones is produced partly by proper cutting and polishing; however, there is also a high correlation between brilliance (in terms of reflectivity) and RI. High refractive index gemstones are capable of fine brilliance when properly cut. For this reason diamond, with an RI of 2.42, displays a high brilliance or reflectivity. Demantoid garnet exhibits one of the highest refractive indices for coloured gemstones. Manson and Stockton disclosed RI values between 1.880 and 1.883 for their twenty-one stones; Payne reported figures to 1.888 for the dark green material. The potential for brilliance is quite outstanding with such high indices.

The specific gravity ranges in Manson and Stockton's study varied narrowly, between 3.80 to 3.88. Payne's stones were somewhat lower, from 3.77 to 3.85.

Another feature of demantoid is the high dispersion that the gem reveals. While most garnets exhibit values in the low moderate range of 0.022 to 0.028, compared to diamond at 0.044. In those low ranges the garnets show very little of the 'fire' which is caused by the white light breaking into its spectral colours. Furthermore, colour masks the effect of the dispersion, and consequently most garnets do not depend on dispersion for their beauty. With demantoid, however, it is a different story. The dispersive powers of demantoid excel those of diamond, with dispersion measurements for the variety at 0.057. Even the colour fails to mask the dispersion completely, especially in the lighter coloured samples.

Unfortunately, demantoid is one of the softest garnets. Its hardness is only rated at 6½ on Mohs scale. If the gemstone is to be mounted in jewellery, then the mounting should be recessed and protected. Or, the gem should be worn as a brooch or necklace rather than a ringstone.

The disadvantage with size is also apparent; there are few stones that exceed one or two carats in size. Gill (1978) reported very rare sizes from 6 carats to 15 carats, owned by dealers or museums. Often they can be found as accessory stones in sizes below 20 points, particularly in antique or old jewellery.

Inclusions in demantoid

The inclusions in demantoid garnet are very distinctive for the variety. Byssolite needle-like fibres are very often found. They usually occur in bundles and radiate from a central crystal of chromite in a curved manner, similar to a horse's tail. These lovely 'horse-tail' inclusions are found in many samples and from various sources of the material.

They are so attractive that the cut stones are often oriented to reveal the byssolite spray in the table area. These inclusions may also be found in a tangled and disordered manner within the crystal. If they occur in straight bundles they can create a cat's-eye (a ¾ carat cat's-eye andradite is known; see *Gems & Gemology*, Fall, 1960; also see discussion on page 108).

Demantoid absorption spectra

The absorption spectrum for demantoid is distinctive. A very strong absorption band is seen centring about 443 nm, seen as a cutoff, with two additional bands at 622 and 640 and a doublet at 693 and 701 (Anderson and Payne, 1955). The first band at 443 was attributed to iron (Fe^{3+}), while the rest were chromium (Cr^{3+}). The demantoid in the San Benito deposit, however, revealed the two chromium bands at 622 and 640, and one narrow band at 685 (perhaps combining the two), but a broad absorption occurred in the violet through to 520 (Payne, 1981, figure 5).

Demantoid sources

Demantoid was first discovered in the Ural Mountains, in veins of chrysolite or in serpentine. The first deposit was at Nizhni-Tagilsk, in the gold washings of the area. The stones were greenish-white to near colourless pebbles. Subsequently, it was discovered in the Sissersk district of the western slopes of the Urals in a stream called Bobrovka (Church, 1879).

First the alluvial material was mined and then another deposit was discovered in the mother rock of the area. The prevailing colour of the early deposits was a yellowish-green, although occasionally, chrome-rich samples were found in the site (Bauer, 1905). Sizes reported were from ¼ to 2 in diameter.

Until Bauer's time (ca. 1896), only one source of demantoid was known, although Miers (1902) reported some green octahedra from the Ala valley in Piedmont (Italy), a known source of topazolite. Indeed, a large number of the yellowish-green demantoids in Manson and Stockton's study were thought to be from this source. But another isolated source of 'emerald-green' demantoid was reportedly Dobaschau (Dobsina) in Czechoslovakia (Dana, 1932, p. 596).

One of the most interesting sources of the gem has been the site reported in San Benito County, California. The initial discoveries of andradite in the area took place in 1949, but in 1979 a mining concern initiated a concentrated effort to mine the area. As many as ten mining claims were taken out in an area surrounded by private property. When Payne wrote his paper on the site (1981), only five demantoid garnets were recovered. Moreover, they were small; they only weighed to ¼ carat for the gem quality crystals, although an additional five stones were found which were cuttable as cabochons, varying in transparency from translucent to opaque. These were slightly larger, weighing up to 1 carat. The site was also important for providing yellow-green andradites and cat's-eye andradites (see discussion below).

Other recent worldwide sources for demantoid have been Korea, the Congo and the Stanley Buttes area of Arizona. In the Arizona site, gems of 2 carats are known, but the colour is described as brownish-green or greyish-green (Sinkankas, 1959, p. 291). The Korean material is very small, producing crystals of only a few millimetres in diameter (Webster, 1983).

Demantoid is a fine gemstone. Its unusually high brilliance and dispersive powers, coupled with an 'emerald-like' green colour and fascinating 'horsetail' inclusions only serve to arouse the interest of gem buyers around the world. Unfortunately, the supply cannot begin to meet the demand.

Yellow andradite (topazolite)

Demantoid often grades into yellow when the chromium element is lacking in strength to produce green colours. Very often demantoid sources produce much more greenish-yellow and yellow andradite than the green demantoid. The yellow colour is attributed to the ferric iron in the andradite, but it may be modified by titanium or manganese as well. Unfortunately, fine pure yellows without the modifying influences of either the green or the brown are unusually rare.

Most of the andradite (35 stones) from the San Benito County area are described as pale yellow-green stones. While only two stones were listed as 'yellow', an interesting phenomenon stone was reported. This was a cat's-eye andradite whose inclusions produced needles which were either coarse or fine. The needles were found in enough profusion occasionally to produce a cat's-eye. A full twenty stones were found in this deposit with this potential, weighing up to 3 carats. Their colours were reportedly yellowish-brown.

Another interesting phenomenon, core zoning, was also found in some stones. This colour was found in a section of the stone near the core. The exterior was yellow in colour while the central core was green. Although the site produced only small quantities of the demantoid and andradite samples at the time of the report (Payne, 1981), nevertheless, it is an exciting source of gem andradite which may lead to future discoveries of larger quantities.

Although there are no reported gem specimens, there does seem to be a miscibility relationship between andradite and spessartite. Deer, Howie and Zussman (1962) list one stone (*Table 8.1*), which reveals 27.8% of spessartite. Fermor suggested the name 'spandite' for such intermediates between the two end-members as early as 1909 (Sastri, 1963).

TABLE 8.1. Unusual occurrence of an intermediate spessartite-andradite from Sweden: named 'spandite' by Fermor as early as 1909

RI	*1.893*
SG	*3.98*
End member composition (mole %)	
Andradite	70.4
Spessartite	27.8
Almandine	1.2
Pyrope	0.4
Grossular	–
Hydroandradite	0.1

Source:
Dark apple green andradite-spessartite garnet, from the Pajsberg manganese and iron mine, Vermland, Sweden; Deer, Howie and Zussman (1962); recalculated by Rickwood (1968)

Uvarovite and the rare garnets

Uvarovite is a chrome-rich green garnet which is named after Count S.S. Uvarov, one-time President of the St. Petersburg Academy in Leningrad. It is the third of the green garnets and it occurs in bright green crystals of a gemmy nature. Unfortunately, the crystals occur in such small sizes that they are not considered a gem material. Furthermore, the rarity of uvarovite makes it difficult to find enough samples to study its unusual composition, not to speak of samples suitable for gemstone use.

If suitable specimens could be found, it would produce an exciting gemstone. It is a calcium-chromium silicate in which chromium, the colouring agent, occurs as a basic ingredient of its bulk chemistry (Cr_2O_3). This is quite unlike demantoid, which owes its rich green colouration to trace elements of chromium. Furthermore, uvarovite's hardness is 7½ on Moh's scale, representing one of the hardest garnets. The RI properties are also high, generally considered to be 1.87 (Webster, 1983), although samples in *Table 8.2* range between 1.798–1.864. Early studies determined that the uvarovite was not fusible with the blow-pipe (Miers, 1902).

In considering the composition of uvarovite, the samples in the literature indicate a solid solution miscibility between uvarovite and grossular (*Table 8.2*). Although there does not seem to be a similar solubility between uvarovite and andradite in the *Table 8.2* stones, nevertheless, the two end-members are in solid solution with each other. Uvarovite has been synthesized, and experiments have produced a solid solution series between uvarovite and andradite hydrothermally at 890°C (Isaacs, 1965). Unfortunately, Isaacs was not able to produce uvarovite-grossulars using either the dry conditions or the hydrothermal conditions.

TABLE 8.2. Uvarovite composition and RI/SG property values. Note the strong miscibility between uvarovite and grossular. Both grossular and andradite are in solid solution with uvarovite

Sample	1	2	3	4	5	6	7
RI	1.798–1.804	1.821–1.829	1.838	1.85	1.855	1.857	1.864
SG	3.712	3.809	3.418	3.75	3.772	3.772	3.75
End-member percentages:							
Uvarovite	49.1	73.3	73.16	89.6	91.2	89.37	87.04
Grossular	41.2	25.7	17.33	1.5	5.5	7.32	5.20
Andradite	5.9	0.9	–	7.8	1.3	1.27	7.52
Almandine	–	–	5.76	–	–	–	–
Pyrope	2.0	1.0	3.75	0.2	2.1	2.05	0.17
Spessartite	–	0.3	–	0.1	–	–	0.07
Hydrogrossular	1.8	0.5	–	0.8	–	–	–

Sources:
1. Uvarovite from Luikonlahti, Finland; Deer, Howie and Zussman (1962) recalculated by Rickwood (1968)
2. Uvarovite from Outokumpo, Finland; Deer, Howie and Zussman (1962) recalculated by Rickwood (1968)
3. Uvarovite from Bisserk, Ural Mts.; Ford (1915)
4. Uvarovite from Outokumpo, Finland; Deer, Howie and Zussman (1962) recalculated by Rickwood (1968)
5. Uvarovite from Outokumpo, Finland; Deer, Howie and Zussman (1962) recalculated by Rickwood (1968)
6. Uvarovite from Kuusjarvi, Finland; Fleischer (1937)
7. Uvarovite from Outokumpo, Finland; Fleischer (1937)

Uvarovite also commonly occurs with trace elements of iron (Fe_2O_3) and titanium (TiO_2) which contribute to some extent to colour alterations (*Table 8.3*, chemical analysis). Frankel (1959) observed that with an increase of iron as a trace element, the uvarovite green becomes 'deeper', and, as the titanium increases, the green colour becomes a 'rusty' green.

The specific gravity of the samples ranged from 3.712 to 3.809 (*Table 8.2*). Studies have shown that there is a relationship between the SG (and the RI) and the chromium content. As the chromium content is increased, the RI, SG (and the unit cell measurement) also increase proportionately (von Knorring, 1951; Frankel, 1959). From the stones of *Table 8.2*, five are presented in *Figure 8.1*, showing this RI/chromium relationship. Like andradite, the pyralspite stones have only a limited miscibility with uvarovite.

Uvarovite is associated with chromite, serpentinized olivine, metamorphic limestones, skarn ore bodies, zoisite and sometimes with diopside, but always dependent upon the existence of chromium in sufficient quantities for its formation. Its original source was Russia, in the Urals near Bisserk, but other sources were soon discovered in Finland, in the Bushveld Complex of South Africa, in Orford, Canada as well as in the Himalayas, Pyrenees, Jordansmuhl in Europe, and California.

In addition to the six major garnet end-members discussed above, there are many other garnets which have been analysed, revealing additional end-member

TABLE 8.3. Chemical analyses of uvarovites, showing bulk and trace chemistry. From Frankel (1959)

	Uvarovite 1	2*	3*	4	5	6	7	8	9	10
RI	1.783–1.786	1.817–1.820	1.824	1.833	1.830	1.801	1.832	1.828–1.830	1.830–1.833	1.837
SG	3.709	3.742	3.763	3.772	3.754	3.740	3.772	3.754	3.789	3.798
Chem. Analysis										
CaO	33.97	–	–	33.70	33.26	35.93	35.12	33.11	33.07	32.81
MgO	1.51	–	–	1.67	6.78	0.62	1.28	–	–	1.86
MnO	–	–	–	0.30	n.d.	n.d.	0.02	–	–	0.03
FeO	1.96	2.70	–	2.73	1.72	2.92	2.51	2.55	2.34	2.53
Cr_2O_3	3.93	9.44	10.56	11.16	11.49	11.54	11.70	11.96	13.64	14.83
Fe_2O_3	4.94	4.86	9.38	6.13	6.30	2.22	6.26	3.96	5.97	3.80
Al_2O_3	n.d.	–	–	9.12	9.44	8.98	7.50	n.d.	n.d.	8.05
$Ti\,O_2$	0.38	n.d.	0.30	1.97	2.24	0.40	2.50	1.31	0.24	1.68
$Si\,O_2$	38.67	–	–	33.50	34.64	36.86	33.33	35.18	35.69	34.52

* partial analysis only

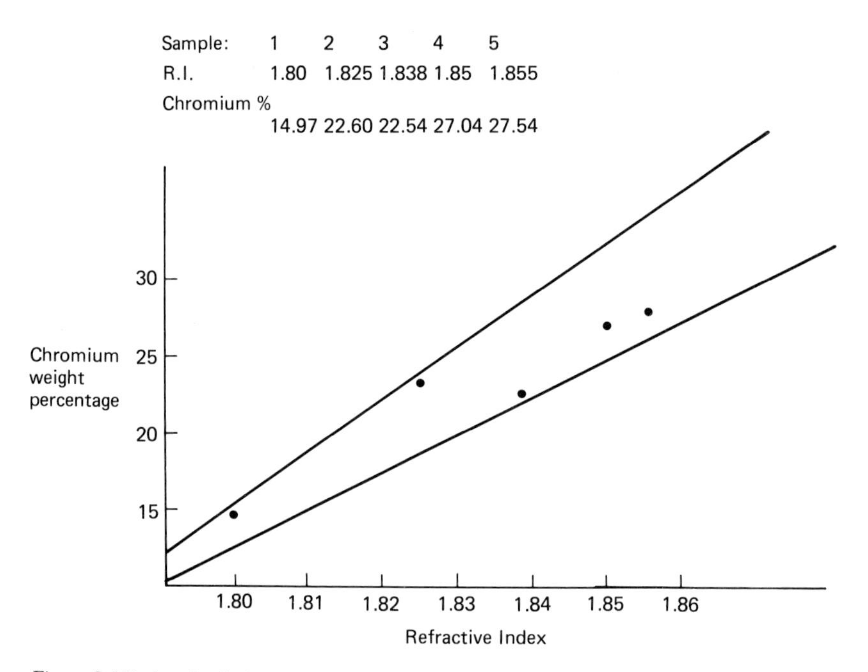

Figure 8.1 Refractive index/chromium relationship in stones 1–5 from *Table 8.2*

possibilities. Although these stones are extremely rare, usually very small in size and not of gem quality, they are explored briefly for a more complete understanding of the complexity of the garnet family. Moreover, the possibility of future gem deposits may exist for some of these species.

These garnets can best be described in terms of their chemistry (*Table 8.4*). There are the magnesium garnets (khoharite and knorringite), the manganese garnets (yamotoite, calderite and blythite), the calcium garnets (goldmanite, kimzeyite, ferric kimzeyite and hydroandradite), the iron garnet (skiagite) and the yttrium garnet (yttrogarnet).

TABLE 8.4. Rare garnet end-members and their properties

Type	*Chemistry*	*Properties*	
		RI	*SG*
Goldmanite	$Ca_3 V_2 (Si O_4)_3$	1.821	3.74
Yamatoite	$Mn_3 V_2 (Si O_4)_3$	1.884	–
Khoharite	$Mg_3 Fe_2 (Si O_4)_3$	1.905	3.885
Knorringite	$Mg_3 Cr_2 (Si O_4)_3$	1.875	3.835
Skiagite	$Fe_3 Fe_2 (Si O_4)_3$	2.010	4.05
Calderite	$Mn_3 Fe_2 (Si O_4)_3$	1.985	4.461
Blythite	$Mn_3 Mn_2 (Si O_4)_3$	–	4.413
Kimzeyite	$Ca_3 Zr_2 (Al_2 Si) O_{12}$	–	3.960
Ferric Kimseyite	$Ca_3 Zr_2 (Fe_2 Si) O_{12}$	–	4.213
Hydroandradite	$Ca_3 Fe_2 H_{12} O_{12}$	1.710	2.784
Yttrogarnet	$Y_3 Al_2 Al_3 O_{12}$	1.823	4.560

Khoharite was discovered and analysed by Fermor (1938). It was found to be a magnesium-iron silicate ($Mg_3Fe_2(SiO_4)_3$ with an RI of 1.905 and SG of 3.885 (McConnell, 1966). Knorringite, on the other hand, is a magnesium-chromium garnet which has taken the place of hanleite, a material thought to be a magnesium-chromium garnet by Fermor (1952). Hanleite, however, was later discovered to be uvarovite and the name was dropped from the nomenclature (Sastri, 1963).

Knorringite was named after Oleg von Knorring of the Department of Earth Sciences, Leeds University, Leeds, England and duly approved by the Commission on New Minerals and Mineral Names of the IMA in 1968 (Nixon and Hornung, 1968). The material was found in a kimberlite pipe in Kao, Lesotho and was a bluish-green colour with a refractive index of 1.803 and specific gravity of 3.756; the Cr weight percentage was 17.47, translating to a 52.7% knorringite end-member percentage if uvarovite is omitted from the casting list, and even a dominant 33.9% percentage of knorringite if uvarovite is retained (*Table 8.5*). The chemical data suggest that the sample is a uvarovite-bearing knorringite (Nixon and Hornung, 1968).

TABLE 8.5. Analysis and end-member percentages of knorringite, the magnesium-chromium garnet found in a kimberlite pipe in Kao, Lesotho

		Number of atoms (0 = 96) in garnet unit cell		
SiO_2	39.92	O	96	
TiO_2	0.11	Si	24.07	24.11
Al_2O_3	9.74	Ti	.04	
Cr_2O_3	17.47	Al	6.91	
Fe_2O_3	1.20	Cr	8.33	15.78
FeO	6.53	Fe^{3+}	.54	
MnO	0.60	Fe^{2+}	3.30	
MgO	16.97	Mn	.33	24.12
CaO	8.14	Mg	15.24	
	100.68	Ca	5.25	

End-member percentage without uvarovite			*End-member percentage with uvarovite*	
Sp.	1.4	RI 1.803	Sp.	1.4
And.	3.4	SG 3.756	And.	3.4
Knorr.	52.7		Knorr.	33.9
Gr.	18.7		Uvarovite	18.7
Alm.	13.9		Alm.	13.9
Pyr.	9.9		Pyr.	28.7

In the manganese garnets (yamatoite, calderite and blythite), the aluminum is replaced with either vanadium, iron or manganese, respectively. The manganese-vanadium garnet, yamatoite, is named after the Yamato Mine in the Kagoshima Prefecture, Japan; however, the IMA rejected the name because the specimen failed to contain at least 50% of the end-member molecule for which the name was proposed (Fleischer, 1965).

Calderite is the manganese-iron silicate which was named originally in 1851 from a rock analysed by Piddington. The specimen was from Bihar in India and was later recalculated by Fermor in 1909. Also, two garnets were reported to be 54% and 35% calderite (weight percentage) from a source in Africa (Sastri, 1963). The specimen studied by Sastri proved to be a spessartite-calderite mixture, with 75.72% molecular percentage of spessartite and 17.85% of calderite (and 5.93% andradite and 0.50% almandine). The spessartite-calderite was an orange-yellow colour garnet from Madhya Pradesh, India.

Blythite was reported by Fermor in 1926. It is the manganese garnet in which the alumina is replaced by manganese, producing the formula $Mn_3Mn_2(SiO_4)_3$. A calderite-blythite mixture was reported by Fermor in 1934 from Chargaon, Madhya Pradesh, India (Sastri, 1963).

The calcium garnets present an interesting variety of combinations. Goldmanite is the calcium-vanadium silicate, the kimzeyites are calcium-zirconium garnets, and the hydroandradite (calcium-iron) is the analogue to hydrogrossular. Goldmanite was first discovered in Laguna, New Mexico and reported in the literature in 1964 (Moench and Meyrowitz, 1964). It was named after Marcus I. Goldman, sedimentary petrologist of the US Geological Service. Sizes of samples were very small, measuring only to 0.10 mm. However, they were free of impurities, unzoned and with a high refractive index, 1.834 (SG 3.765). Although the material is found in association with uranium, no traces of uranium were found in the garnet. Chemical analysis (*Table 8.6*) reveals a goldmanite-grossular miscibility with a high vanadium weight percentage (18.3%). The material was also synthesized in 1965 (Strens, 1965).

TABLE 8.6. Analysis and composition of goldmanite, the calcium-vanadium garnet from Laguna, New Mexico. From Moench and Meyrowitz (1964)

		Chemical analysis	
RI	1.821		
SG	3.74	SiO_2	36.6
End-member composition		Al_2O_3	4.9
		Fe_2O_3	5.4
Goldmanite	59.8	V_2O_3	18.3
Grossular	20.1	CaO	33.3
Andradite	16.7	MgO	0.7
Pyrope	2.7	MnO	0.3
Spessartite	0.7	*Total*	99.5
Almandine	–		

Kimzeyite was named in 1961, but a ferric-kimzeyite was proposed to account for all of the available zirconium. Both kimzeyites have been synthesized by Ito and Frondel (1967). Natural zirconium-rich titanium garnets have been found in the Marathon Dikes near Marathon, N.W. Ontario (Platt and Mitchell, 1979). Sizes of the stones range to below 0.2 mm and their colours were reported pale reddish-orange. End-member components involving Mg-melanite, Fe^{2+}-melanite, schorlomite, Al-kimzeyite, Fe^{3+}-kimzeyite and andradite were necessary to describe these unusual stones. The ferric-kimzeyite ($Ca_3 Zr_2 Fe_2^{3+} Si O_{12}$) was found in molecule percentages from 4.55% to 28.35%; the Al-kimzeyite molecule ($Ca_3 Zr_2 Al_2 Si O_{12}$) ranged from 1.10 to 15.75%.

Hydroandradite has been synthesized in 1941 (Rickwood, 1968). The RI was reported as 1.710 and the density at 2.784. Yttrium-rich garnets (yttrogarnets) reveal a high refractive index (1.823) and density of 4.560 (Yoder and Keith, 1951). Skiagite ($Fe_3 Fe_2 Si_3 O_{12}$) was reported by Fermor (1926) with high properties also, at 2.010 (RI) and 4.567 (SG) (McConnell, 1966).

In addition to these rare garnets, future synthesis research and/or natural garnet discoveries could uncover more possibilities. However, the rare garnets described above illustrate both the complexity of the garnet composition and the variety of end-members within the garnet family. Occasionally, an understanding of these rare specimens will be necessary to explain the composition of new natural garnets, as illustrated in the study by Platt and Mitchell (1979). Whether or not these rare end-members will ever occur in gem grade specimens is unknown. Nevertheless, as long as the possibility exists, and new discoveries of garnets are conceivable in exotic combinations, knowledge of their existence seems important enough to include in a study of gem garnets.

Colour-change garnets

In 1970 there was a report of a very unusual gem grade garnet which was discovered in Tanzania. The stone was a rolled pebble of 3½ carats and only slightly included with needle-like crystals. The colour of the stone was reported to be quite unusual by the author (Crowningshield, 1970). In daylight the stone appeared blue-green, and under incandescent lighting, it was a purple-red. X-ray diffraction tests were conducted on the uncut sample, confirming its identity and an analysis was carried out by Pacific Spectrochemical Laboratory, revealing that the garnet was manganese-magnesium silicate.

In a later study (Jobbins *et al.*, 1975), the same stone was calculated for end-member percentages. It was a pyrope (53.7%)-spessartite (39.0%) with some andradite (4.5%), uvarovite (1.7%) and goldmanite (1.1%). The RI/SG of the stone was 1.765 and 3.88, respectively. The absorption spectrum of the sample was also presented. The cause of the colour-change was speculated to be due to vanadium, and possibly chromium.

In the study of Jobbins *et al.* (1975), the vanadium rôle was considered to be the dominant cause of the colour-change, particularly since the stone they studied seemed to be lacking in chromium. However, in presenting the absorption spectrum, there was a hazy band about 50 nm wide, from about 550 to 600 nm, centring at about 575 nm. The existence of this band was puzzling, since chromium was not thought to be present in the specimen. Nevertheless, a definite similarity was apparent in the absorption pattern of the Crowningshield (1970) study.

Another study of colour change garnets presented data on two more East African stones (Stockton, 1982). At first glance, the colour-change seemed less pronounced, changing only from reddish-orange to purplish-red in general. However, the colour-change was also noticed from transmitted light to reflected light. Under transmitted fluorescent lighting, stone A displayed a greenish-yellow brown; under reflected fluorescent lighting, the colour observed was a purplish-red. Stone B, under the same lighting, changed from light bluish-green to purple. Under incandescent lighting, however, the change from transmitted to reflected light varied from reddish-orange to red for stone A; stone B varied from red to purplish under the same lighting sources. Stone A proved to be 59.02% (molecular percent) spessartite, 20.33% andradite. Stone B was a spessartite-pyrope intermediate with less than 50% of the spessartite end-member (43.16%).

The absorption spectrum of the two stones proved interesting, however. Instead of two high peak areas of transmission noticed in the first two studies, only one high

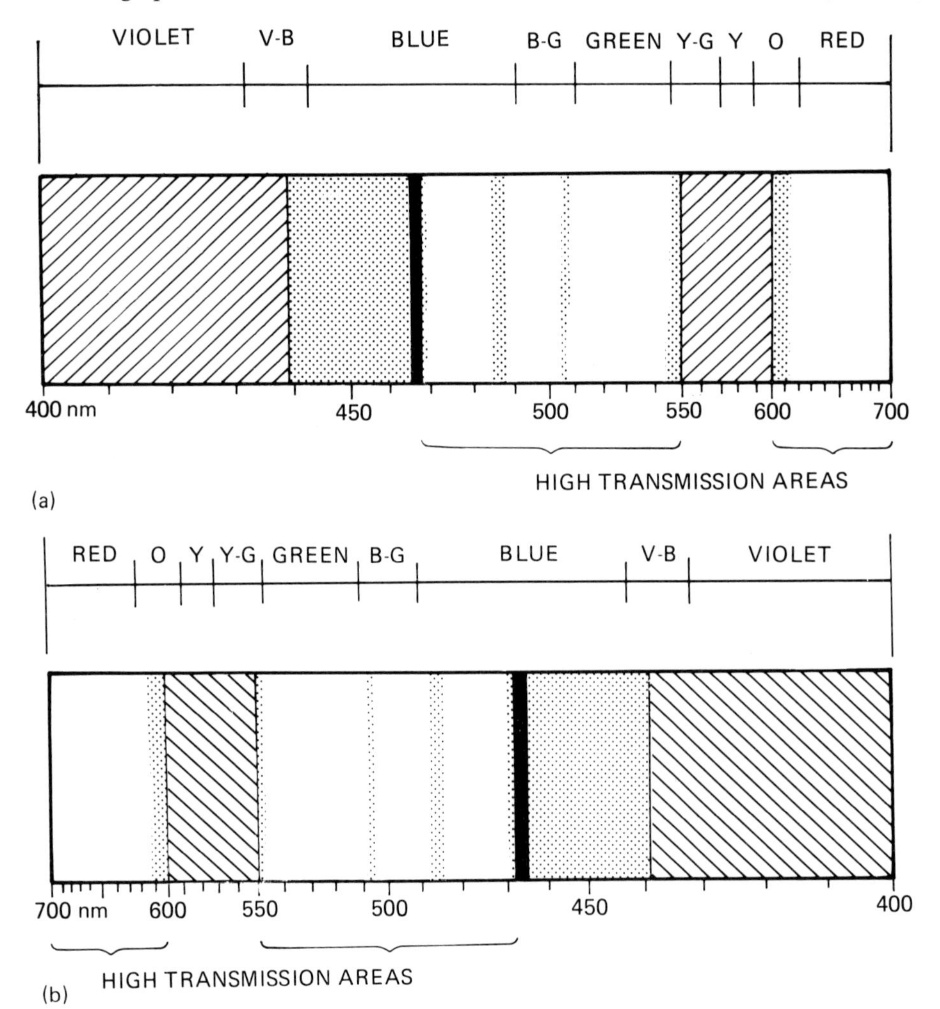

Figure 8.2 Absorption spectrum of colour-change garnet after Crowningshield, 1970. High transmission peaks in the blue/green and the red account for the colour change (Gübelin and Schmetzer, 1982) (a) USA; (b) Europe

transmission peak occurred in the orange to red area of the spectrum. There was no secondary transmission found in the blue-green area of the spectrum, although stone B did reveal a weak transmission hump, centring about 500 nm. This difference suggests a colour change of a different nature than the earlier studies.

A comparison of the two stones under reflected light, reveals that the colour-change is less significant, from purplish-red to purple (stone A) and purple to purplish-red. Such a change perhaps should be attributed to the different spectral distribution emitted from the two light sources. Office fluorescent lights emit energy peaks high in the blue to green area of the spectrum, while incandescent lighting produces an overbalanced energy in the red area of the spectrum. The comparison of suspected colour-range material is better studied under another lighting source, perhaps one with a grading lamp which presents a closer approximation to ideal daylight.

Another study of the colour change gemstones in general presented data on several gemstones, including garnet (Gübelin and Schmetzer, 1982). In this study it was concluded that cause of the colour-change was not to be found 'in the crystal structure or in the chemistry of the colouring agents, but rather in an equivalent correspondence of the position of the absorption maxima and minima in the visible region of the spectrum'.

In particular, the high peaks of transmission in the two areas of blue-green to violet and the orange to violet were especially noticed as a trend in the colour-change materials. These transmission areas (*Figure 8.2*) are also seen in the stones reported by Crowningshield (1970) and by Jobbins *et al.* (1975), but not in the stones reported by Stockton (1982), and are consistent with our observations above. Consequently, light sources with unbalanced spectral distribution patterns, as evident in both fluorescent and incandescent light sources, react with the strong selective absorption pattern of the gemstones to create the colour-change phenomenon.

In conclusion, the garnet group of gemstones and minerals presents and intriguing and fascinating array of rarities and unusual phenomena. A stone of unusual interest can be seen in the asteriated rhodolite garnet from East Africa, illustrated in *Plate 21*. Under reflected light, the star exhibits four rays. However, when the light is transmitted underneath the stone ('diasterism') a six-ray star appears. Along the side of the stone, another six-ray star is seen by transmitted light. When coupled with the almandine star garnets which present four-ray stars with six-ray stars at the end of each leg of the main star (described in Chapter 5 Almandine), the asteriated garnets present fascinating specimens of unusual interest.

Summary of andradite properties

Chemistry:	$Ca_3\,Fe_2\,(Si\,O_4)_3$
Colouring agents:	Fine green: chromium in trace amounts
	Other colours: ferric iron
Refractive index:	1.880–1.888
Specific gravity:	3.77–3.88

Absorption spectra:	Strong band at 443 nm, appearing as a cutoff 622 nm and 640 nm 693 nm and 701 nm (doublet)
Hardness:	6½
Colours:	Demantoid: Green Topazolite: Greenish-yellow to yellow
Dispersion:	0.057
Typical inclusions:	'Horsetail' pattern of byssolite fibres

Bibliography

ANDERSON, B. W. and PAYNE, C. J., 'The absorption spectrum of the demantoid garnet', *The Gemmologist* (January, 1955)

BANK, HERMANN, 'Demantoid garnet from Korea', *Gems & Gemology* (Winter, 1978–1979)

'Cat's-eye Demantoid', *Gems & Gemology,* Fall (1960)

CHURCH, A. H., 'On the so-called green garnets, from the Urals', *Mineralogical Magazine,* **2** (1879)

CROWNINGSHIELD, ROBERT, 'A rare alexandrite garnet from Tanzania', *Gems & Gemology* (Summer, 1970)

FERMOR, L. L., 'On the composition of some Indian garnets', *Records of the Geological Survey of India,* **73** (1926)

FERMOR, L. L., 'Garnets and their role in nature', Calcutta (1938)

FERMOR, L. L., 'On khoharite, a new garnet and on the nomenclature of garnets', *Records of the Geological Survey of India,* **73,** pt. 1 (1938)

FERMOR, L. L., 'On a new chrome-garnet', *Geological Magazine,* **89** (March-April, 1952)

FLEISCHER, MICHAEL, 'New mineral names', *The American Mineralogist,* **50** (1965)

FRANKEL, J. J., 'Uvarovite garnet and South African Jade (hydrogrossular) from the Bushveld Complex, Transvaal', *The American Mineralogist,* **44** (May-June, 1959)

GILL, JOSEPH O., 'Demantoid – the complete story', *Lapidary Journal* (October, 1978)

GÜBELIN, E. and SCHMETZER, K., 'Gemstones with alexandrite effect', *Gems & Gemology* (Winter, 1982)

ISAACS, T., 'A study of uvarovite', *Mineralogical Magazine,* **35** (1965)

ITO, J. and FRONDEL, C., 'Synthetic zirconium and titanium garnets', *The American Mineralogist,* **52** (1967)

JOBBINS, E. A., SAUL, J. M., TRESHAM, ANNE E. and YOUNG, B. R., 'Blue colour-change gem garnet from East Africa', *The Journal of Gemmology,* **XIV,** No. 5 (January, 1975)

KNORRING, VON, OLEG, 'A new occurrence of uvarovite from northern Karelia in Finland', *Mineralogical Magazine,* **29** (1951)

MANSON, D. VINCENT and STOCKTON, CAROL, 'Gem andradite garnets', *Gems & Gemology* (Winter, 1983)

MANSON, D. VINCENT and STOCKTON, CAROL, '"Fine green" demantoids', *Gems & Gemology,* **XX,** No. 3, Fall (1984)

MCCONNELL, D., 'Refringence of garnets and hydrogarnets', *Canadian Mineralogist,* **8** (1964)

MIERS, HENRY A., *Mineralogy* (1902)

MOENCH, ROBERT H. and MEYROWITZ, ROBERT, 'Goldmanite, a vanadium garnet from Laguna, New Mexico', *The American Mineralogist,* **49** (May-June, 1964)

MOHS, FREDERICK, *Treatise on Mineralogy,* **II** (1825)

NIXON, PETER H. and HORNUNG, GEORGE, 'A new chromium garnet end member, knorringite, from kimberlite', *The American Mineralogist,* **53** (November-December, 1968)

PAYNE, TEDD, 'The andradites of San Benito County, California', *Gems & Gemology,* Fall (1981)

PLATT, R. GARTH, 'The Marathon Dikes. I: zirconium-rich titanium garnets and manganoan magnesian ulvospinel-magnetite spinels', *The American Mineralogist,* **64** (1979)

RICKWOOD, P. C., 'On recasting analyses of garnet into end-member molecules', *Contr. Mineral. and Petrol.,* **18** (1968)

SASTRI, G. G. K., 'Note on a chrome and two manganese garnets from India', *Mineralogical Magazine,* **33** (June, 1963)

STOCKTON, CAROL M., 'Two notable color-change garnets', *Gems & Gemology* (Summer, 1982)

STRENS, R. G. J., 'Synthesis and properties of calcium vanadium garnet (goldmanite)', *The American Mineralogist,* **50** (January-February, 1965)

VERMAAS, F. H. S., Manganese-iron garnet from Otjosondu, South-West Africa', *Mineralogical Magazine,* **129** (1952)

YODER, H. S. and KEITH, M. L., 'Complete substitution of aluminum for silicon: the system $3MnO \cdot Al_2O_3 \cdot 3SiO_2$-$3Y_2O_3 \cdot 5Al_2O_3$', *The American Mineralogist,* **36** (1951)

Appendix 1: Colour analysis of 'Pyrandines' from the Malagasy Republic and Orissa

During the autumn of 1983 and the spring of 1984 there was an opportunity to analyze the properties and colours of two groups of garnets. One group was from the Malagasy Republic, and the other was from Orissa, in India. A total of sixty-three Malagasy gems and twelve Orissa garnets comprised the two groups. Gemmological tests performed on the stones included optical, density and inclusion studies. Three separate methods for describing colour were also used, including a visual perception method, the ColorMaster with the GLA Color Grading Manual, and a new gem spectrophotometer.

In reviewing the literature, the first article on Malagasy garnet was a brief study of its properties by Tisdall in 1962. In his analysis one stone was selected from a parcel and tests were performed, including an unusual magnetic test originated by Anderson (1959). The revealed an RI of 1.762 and SG of 3.84, nearly identical to the properties reported reported from the North Carolina rhodolites, as reported by Ford in 1915 (RI 1.7596; SG 3.837). The colour of Tisdall's stone was described as purplish-red, also consistent with the North Carolina colours.

In another study of Malagasy garnets, ten gemstones were reported to range from 1.750 to 1.755 (RI), with an SG from 3.81 to 3.86 (Campbell, 1973). In this study also, the colours were described as 'identical' to the North Carolina rhodolites (violetish-red).

From the above studies it was expected that the Malagasy gemstones would fall into the rhodolite classification (RI between 1.750–1.780) and exhibit colours on the violet side of red. The Orissa stones were not studied previously.

Gemmological findings

The RI analysis seemed to indicate a homogeneous grouping from both sources. No spessartites or almandines were found in the sample tested. Both groups were typically too high in properties to be classified as pyropes, and too low to be called almandine. They ranged from 1.749 to 1.769 for the Malagasy stones, and 1.749 to 1.759 for Orissa. Refractive indices were taken with a Gem Duplex refractometer using monochromatic lighting.

Specific gravity tests were conducted hydrostatically, using a Scientech model 222 electric balance. Weights were conducted in air and water. The specific gravity data for both groups followed the pattern of the refractive indices. The Malagasy stones ranged slightly lower than the Orissa gems, at 3.82 to 3.90 and 3.73 to 3.86, respectively. Malagasy averaged at 3.85 and Orissa slightly lower, at 3.78.

Spectroscopic tests were conducted on both groups of garnets with an Ealing prism spectroscope. Almandine lines were easily seen throughout both groups, at 505, 525 and 575 nm, characteristic of the pyrope-almandine intermediates. Also, faint lines were sometimes seen at 428 and 617 nm. There were no distinct differences between the Madagascar and Orissa garnets in this respect.

Inclusions were found in both groups of stones, although the Orissa garnets were much less included than the Malagasy gems. Several types of inclusions were found in the Malagasy stones. Crystal inclusions were of two types: well-rounded crystals with no significant habit and those revealing a hexagonal form. The latter crystals might have been rutile and/or apatite crystals which are common for such garnets (Zwaan, 1974, etc.). The rounded crystals appeared to be opaque and were quite common, occurring in nine of the eighteen samples. Some of the Madagascar stones exhibited a profusion of such inclusions, particularly in sample number two, which also revealed a considerable strain due their presence. The hexagonal crystal was rare and occurred with rounded edges along the length of the crystal.

Needle-like inclusions were very common in the Malagasy garnets. They varied in their orientation from one or two directions to many directions, sometimes intersecting in angles of 70° and 110°. They also varied from long to short. In addition to the needles, many Madagascar stones displayed very tiny clusters of 'dust-grain' inclusions. Unlike the stones in Campbell's study (1973), no two-phase or three-phase inclusions were found. However, in one Madagascar stone, there was a very unusual fingerprint inclusion. The compactness and appearance of the fingerprint closely resembled corundum fingerprint-type inclusions. Trumper (1952) photographed a rare fingerprint in a Ceylon rhodolite, so this phenomenon cannot be considered unique to the source.

The Orissa gemstones were remarkably free of inclusions. Virtually one-half of the stones examined were flawless. Among the remaining half, only tiny clusters of 'dust-grain', platy and needle-like inclusions were present. The needles were often short and usually oriented in several directions. No differences were seen in the Malagasy needle-like and 'dust-like' inclusions when compared with those from Orissa. The platy crystals were not found in the Malagasy stones, but they are probably common to garnet (Zwaan, 1974, p. 9). Also, two-phase inclusions were found in one of the Orissa garnets. If it can be concluded that the Orissa stones are closer in properties to the pyrope end-member, then one might reasonably argue that the characteristic clarity one finds in pyrope is also associated with pyrope-almandines which incline toward pyrope. None of the inclusions found in either groups of garnets could be considered unique to the sources.

Colour analysis

In order to arrive at accurate colour definitions of our test stones, it was decided to use three separate methods for describing colour. A visual technique of colour evaluation was one method used. In this system colours are described in terms of hue, tone and intensity and structured around the international colour systems of

Munsell, DIN 6164 and others. The unaided eye was used to determine the colour definitions, a method based on normal colour vision (as tested by the Munsell-Farnsworth 100 hue discrimination exam) and viewing gems under well-balanced grading lamps.

A second method was also visual in application; but it involved the use of GIA's ColorMaster, which was created to discriminate and describe colours in gemstones. The GIA coloured stone grading manual was also used to arrive at hue, tone and saturation levels.

A third method of colour definition was the application of a new gem spectrophotometer, the Gemcolor II by Kalnew in Japan. Although this machine was found to be inadequate for diamond colour analysis because the small spread of spectral data was not adequate for the narrow 'cape-line' in diamonds, it seemed to be very useful in analyzing the spectral power distribution in coloured stones. For example, rubies with increasing amounts of purple in the hue-mix would reveal transmission peaks in the blue area of the spectrum. It was also found that, with practice, the colour appearance of the gem could be estimated accurately just by studying the spectral power distribution curves. However, it should be pointed out that this machine, unlike most spectrophotometers, evaluated gemstones from a face-up viewing position through the use of a patented gem holding device. Light entered through the crown of the stone and then was transferred to the photomultipliers.

Results of the visual perception method

The hue position as determined by the visual perception method was largely orangy-red for the Malagasy stones. One Malagasy gem was estimated to be red; the orangy-red stones varied from slight orange to moderate orange in the hue-mix. In the Orissa stones, only four of the twelve were considered to be orangy-red, while eight were considered to be violetish-red, varying from slight violet to moderate violet in the hue-mix. The tones (lightness to darkness of the hue) of the Malagasy stones were considered to be darker than the Orissa gems. On a scale of 0-100, the zero being colourless and one hundred being black, the Malagasy gems ranged from 60 to 85, while the Orissa gems were estimated from 50 to 75. The intensity (colour saturation) was into the moderate to bright categories for both groups, but failed to achieve the highest, or vivid category.

Results of the ColorMaster system

The hues of the ColorMaster confirmed the visual hue descriptions with very few exceptions. This agreement is not surprising, since the colour match on the Colormaster is accomplished by visual perception. The differences, however, although slight, revealed disagreements on the amounts of orange in the hue-mix of the orangy-red stones. Both tone and intensity estimations derived from the ColorMaster were very close to the results obtained in the visual perception method above.

Results of the Gem Spectrophotometer

The spectrophotometer features a double-beam photometer with an integrating sphere and incorporates a spectroscope with a blazed holographic grating (to identify the cape line in diamonds). The spectroscope itself is interfaced with a microcomputer which complies and records data both graphically and numerically (plotter/printer). The graphic information is displayed from 400 nm to 700 nm in the visible spectrum, with an alternate range into the ultraviolet from 360 nm to 700 nm. Numerical printouts are possible for every 20 nm of the spectral range (the main drawback for diamonds). The numerical data are given in CIE terminology, which includes the spectral tristimulus values (X,Y,Z), chromaticity coordinates in the CIE colour space (x and y values), dominant wavelength (hue position) and excitation purity (saturation).

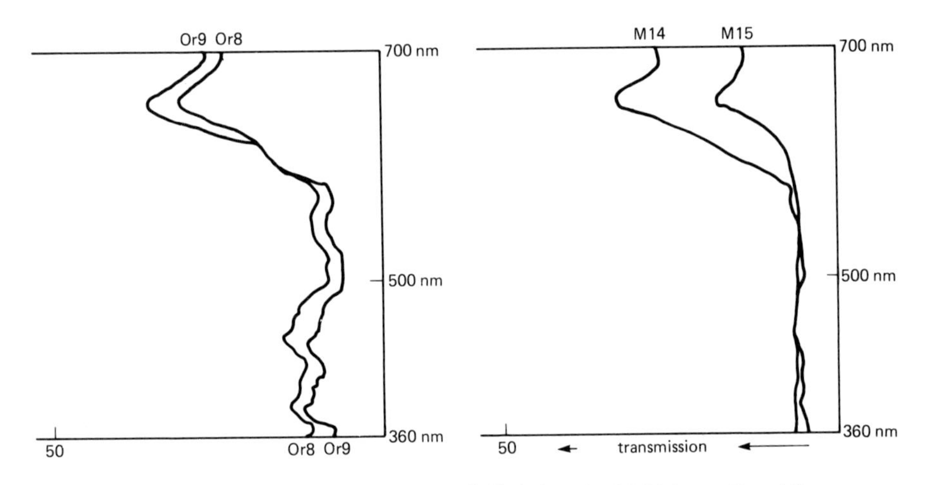

Figure 1 Spectral power distribution comparison (*left*) Orissa; (*right*) Malagasy Republic

When submitting the Malagasy and Orissa garnets to the machine, it was found that the spectral power distribution curves seemed to be characteristic for each source (see *Figure 1* for a comparison of two Orissa gems with two Malagasy stones). Whether or not other coloured stones would exhibit a similar pattern for each source was not studied; however, with a larger group of stones from the same source, it is quite possible that this observation may not hold true, particularly where the source produces a full range of colours.

Hues reported by the spectrophotometer were given in dominant wavelengths according to CIE terminology. Unfortunately, the CIE system was designed to give precise and consistent matching data on colours, and not define them with terms. Although verbal colour terms have been used to define the CIE colorspace as early as 1943, with revisions in 1955, the boundaries drawn are entirely arbitrary, and may not conform to visual observation. The resulting data from the spectrophotometer indicated that all the Malagasy stones except three were into the orangy-red hue position with dominant wavelengths from 608.1 nm to the arbitrary cutoff at 629 nm. The three remaining stones registered dominant wavelengths slightly higher, into the reds. All of the Orissa stones except one were from the slightly purplish-red to purplish-red, with dominant wavelengths in the complementary values from 493 C ("C" for complementary in the non-spectral colours) to 500.7 C. The one Orissa stone that was the exception was graded 'red' at 641.8 nm. This hue information basically supported the visual methods described above, especially with the Malagasy stones. The lack of orangy-red colours in the Orissa stones presented conflicting data, however, with those visual methods which described some Orissa stones as orangy-red. Reasons for the discrepancy probably relate to the colour space boundaries in the CIE system.

Tonal data were derived from the CIE 'Y' value of the tristimulus values X, Y, Z (not to be confused with the lower case letters indicating the CIE color space). The 'Y' value is also called the 'luminance factor' and correlates well with the lightness-darkness of the sample. There were some differences found when comparing the ColorMaster and the visual system; nevertheless, averages of the 'Y' values indicated that the Orissa stones were generally lighter in tone than the Malagasy gems, which was consistent with the visual methods.

Intensity data also revealed inconsistencies. The excitation purity of the CIE system is generally equivalent to saturation, or intensity. Higher values should indicate higher (i.e. more vivid) intensities of the hue. Yet, in comparing Malagasy stone number 15 which exhibited a 10.7% excitation purity, with Malagasy stone number 5 which displayed 30.3% excitation purity, there was no preceivable saturation or intensity difference visible whatever. Moreover, the saturation levels were well below mid-range, and far below other garnets reported in CIE terminology by Manson and Stockton (1981, p. 199), and Stockton and Manson (1982, p. 337). Perhaps such contradictions should empasize the fundamental underlying differences between the CIE system and systems of colour analysis which depends upon the perceived appearance of a colour sample in the world around us. The CIE system functions to tell whether a colour matches, not what it looks like in real life.

Colour analysis conclusions

Colour separations and descriptions have always been difficult in gemstones, especially when the techniques for evaluating colour have not been standardized. The importance for precise colour definition in gemstones is realized when separations are attempted between ruby and pink sapphire, between green beryl and emerald, between green grossular garnet and tsavolite, and between demantoid and greenish-yellow topazolite. The spectrophotometer could be useful in making such distinctions; but from studies with the Malagasy and Orissa garnets, it would seem that long term studies with the machine would be necessary to validate its reliability and usefulness. Furthermore, tests from one instrument to another would be required to ensure consistency.

The visual methods used above for defining gemstone colours seemed adequate to define broad differences between the purplish-red stones and the orangy-red samples. Moreover, the use of the Colormaster would probably achieve more consistency among observers, since few people are trained in the visual system techniques. Nevertheless, both systems relate to colour appearance, and they both were reasonably consistent in defining the hue positions of the samples.

It can be concluded that the Malagasy gemstones in the study were found to be either red or orangy-red. This conclusion was consistent with all three systems for deriving gemstone colours. Earlier reports of Malagasy garnets with a 'rhodolite' colour (purplish or violetish-red) may have represented exceptional colours, or the techniques for arriving at the colour may have relied upon analysis by transmitted light, rather than internally relected light, a technique, incidentally, which produces more violetish-red hue mixes, particularly in uncut gem samples, but also in cut stones. However, because of the possibility of one source producing a full range of colours (i.e. the Orissa stones), it is also conceivable that the Malagasy mines indeed produce colours that could be defined as purplish or violetish-red.

From the gemmological analysis, we would categorize the samples from Malagasy and Orissa as pyrope-almandine intermediates. We would also say that the purplish-red Orissa gemstones could be called 'rhodolite' and the orangy-red stones from either source could be called 'pyrandine' (after Anderson, 1959), based on the colour analysis performed above.

Bibliography

ANDERSON, B. W., 'Pyrandine – a new name for an old garnet', *The Journal of Gemmology,* pp. 15–16 (April 1947)

ANDERSON, B. W. and PAYNE, C. J., 'The spectroscope and its application to gemmology, part 19: absorption spectrum of almandine garnet', *The Gemmologist,* pp. 43–46 (March 1955)

ANDERSON, B. W., 'Properties and classification of individual garnets', *The Journal of Gemmology,* pp. 1–7 (Jan. 1959)

CAMPBELL, IAN C. C., 'A comparative study of Rhodesian rhodolite garnet in relation to other known data and a discussion in relation to a more acceptable name', *The Journal of Gemmology,* pp. 53–64 (April 1972)

CAMPBELL, IAN C. C., 'A gemological report on rhodolite garnet, Malagasy', *Lapidary Journal,* pp. 958–960 (Sept. 1973)

HANNEMAN, W. WILLIAM, 'A new classification for red-to-violet garnets', *Gems and Gemology,* pp. 37–42 (Spring 1983)

HIDDEN, W. E. and PRATT, J. H., 'On rhodolite, a new variety of garnet', *American Journal of Science,* **V,** pp. 294–296 (April 1898)

JOBBINS, E. A., SAUL, J. M., STATHAM, PATRICIA M., and YOUNG, B. R., Studies of a gem garnet suite from the Umba River, Tanzania', *The Journal of Gemmology,* **XVI,** 3, pp. 161–171 (1978)

MANSON, D. VINCENT and STOCKTON, CAROL M., 'Gem garnets in the red-to-violet color range', *Gems and Gemology,* pp. 191–204 (Winter 1981)

MARTIN, B. F., 'A study of rhodolite garnet', *The Journal of Gemmology,* pp. 29–36 (April 1970)

SCHMETZER, KARL and BANK, HERMANN, 'Garnets from Umba Valley, Tanzania: is there a necessity for a new variety name!' *The Journal of Gemmology,* pp. 522–527 (Oct. 1981)

STOCKTON, CAROL M. and MANSON, D. VINCENT, 'Gem garnets: the orange to red, orange colour range', *International Gemological Symposium Proceedings 1982,* edited by Dianne M. Eash, pp. 330–338

TISDALL, F. S. H., 'Tests on Madagascar Garnet', *The Gemmologist,* pp. 102–103 (June 1962)

TRUMPER, L. C., 'Rhodolite and the pyrope almandine series', *The Gemmologist,* pp. 26–30 (Feb. 1952)

ZWAAN, P. C., 'Garnet, corundum and other gem minerals from Umba, Tanzania', *Scripta Geologica,* **20,** pp. 1–41 (1974)

ZWAAN, P. C., 'Solid inlusions in corundum and almandine garnet from Ceylon, identified by X-ray powder photographs', *The Journal of Gemmology,* pp. 224–234 (May 1967)

Appendix 2: Garnet absorption spectra

Almandine (a) The principle iron lines are seen at 505, 525, 575. Additional lines may sometimes be seen at 393, 404, 428, 438, 462 and 476, with another line at 617 nm.

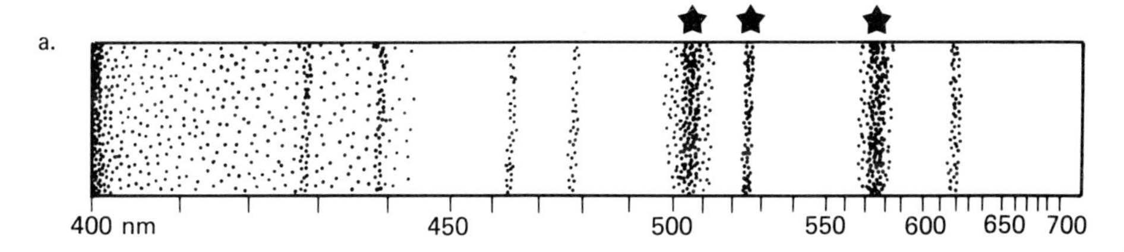

Pyrope (b) Chrome pyrope will show a broad band from about 520 to 620; there will also be a strong line at 505 nm, revealing the iron (ferrous) content in the sample from the almandine end-member.

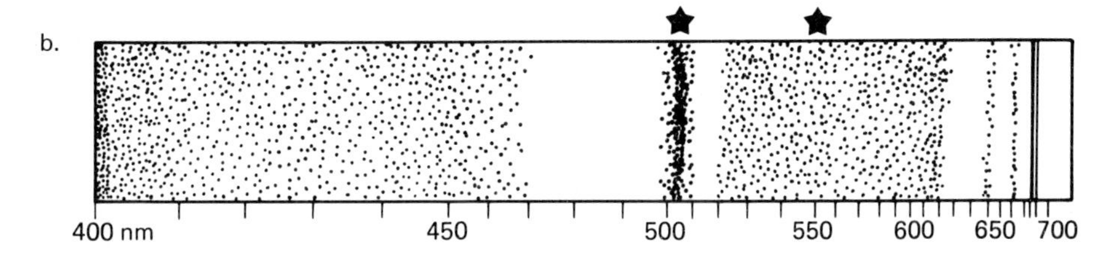

Spessartite (c) The manganese is responsible for a strong band at 432 nm; additional spectra, sometimes difficult to see in the violet end of the spectrum are a strong fine line at 424 and a strong band at 412 nm. Additional lines sometimes seen are weak lines at 462, 485, and 495 nm. Liddicoat (1972) found three additional lines in light orange spessartites from Amelia, occurring as weak bands at 510, 532 and 560 nm.

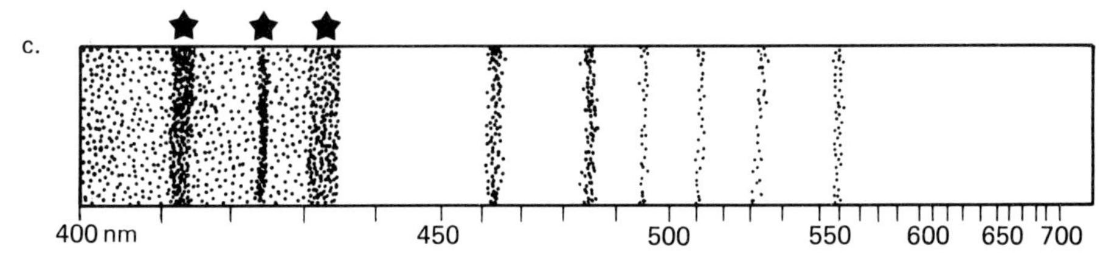

Demantoid (yellowish-green) (d) The Ferric iron is responsible for this spectrum. A strong band centres about 400 nm, while there is some absorption at the end of the spectrum. (after Liddicoat, 1972).

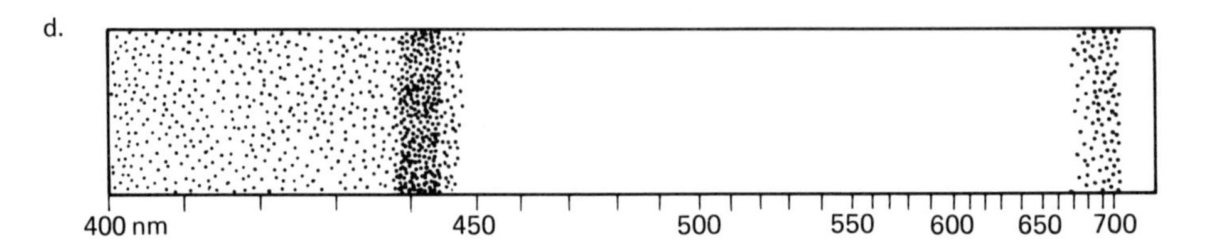

Demantoid (intense-green) (e) There is general absorption to about 475 nm, and two strong narrow bands at 620 and 638 nm. An additional fine line can be seen at 690 nm. (after Liddicoat, 1972).

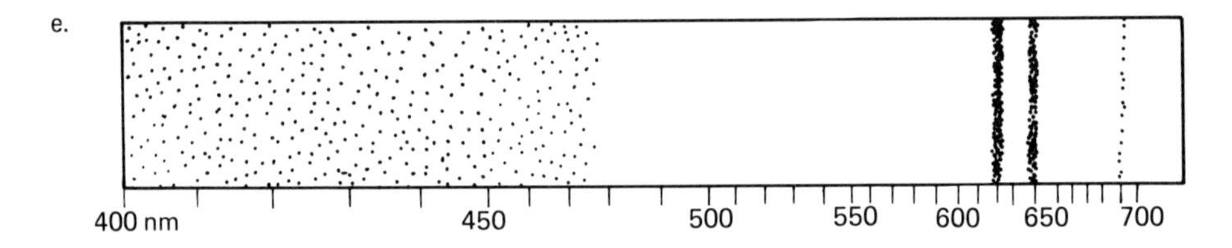

Grossular Absorption spectra is rarely seen in grossular garnets. If it is visible, there may be some resemblance to the spessartite spectra, for a strong band generally occurs in the same vicinity (about 430–440 nm).

Appendix 3: Garnet jewellery in ancient and medieval history

The use of garnets in jewellery extends back into the very earliest periods of human history. The earliest example in the British Museum is a bead diadem dating from the Naqada II period (about 3200 BC). The diadem incorporates very tiny beads, drilled and polished and set with similarly cut and polished malachite and turquoise beads. Most of the beads are garnets, and they are arranged in colourful groups, separated by loops of gold beads. The garnets are not machine cut, for their shape varies with each bead; and, they appear to be orangy-red coloured almandines. The diadem was found at Abydos and is an important item in the Egyptian jewellery collection of the British Museum.

In the later periods of Egyptian history, garnets were rarely used for jewellery. Carnelian, sard and sardonyx beads, however, were very common and many examples can be seen in museums. The beads were commonly strung into necklases rather than diadems in these periods. However, the British Museum has on display at least one garnet necklace, dating from the Middle Kingdom (about 2133–1567 BC). It is completely composed of garnet, and the beads are graduated, well-rounded (like wheels) and well polished. The garnets appear to be nearly opaque, brownish-orange in colour and range in size from very small to about 8 mm diameter. However, garnet was not the most favoured gemstone throughout the Egyptian periods, and few other examples can be found.

Even down to the Classical Greek period (5th century BC), the use of garnets was limited only to intaglios, and they were used then only rarely. But the widespread use of garnets really dates from the 4th century, particularly after Alexander the Great opened the riches of the Orient to Greece and the West. From the 4th century to the 1st century BC, the period of history known as 'the Hellenistic' spread throughout the Mediterranean lands, diffusing much of the eastern cultural achievements into the west. Intaglio seals continued to be made in the tradition of the times. However, a profusion of seals cut in garnet occurred, probably as a direct result of the garnet trade with India opened up by Alexander. Seals cut in the form of gods were now made to look like the god-kings (Alexander) who seemed to achieve miracles with their armies and administrative skills. They were heroes and saviours to the people; it is no wonder that the kings were thereupon commemorated on coins and intaglios as gods among men. An interesting example of this commemoration in a garnet intaglio is seen in *Figure 1*. The intaglio dates from the early Hellenistic period and exhibits the bust of an idealized god-like person with long, flowing hair. The treatment of the hair suggests that this figure is Alexander the Great portrayed as Helios, the sun god complete with radiated head. Intricately carved around the field are tiny symbols of the zodiac. The garnet is brownish-red in colour and probably almandine from India.

Figure 1 (*left*) Garnet intaglio seal stone (18 × 16 mm), with radiate head. Hair treatment is done in the style of the Alexander portraits. The figure is Helios, the sun god surrounded in the margin by carefully executed symbols of the zodiac. The stone is a high-dome cabochon intaglio. The date is Hellenistic, perhaps in the third century BC, or, at the earliest, late fourth century BC. BM gem collection #1168. (*Photo by author; courtesy of the Trustees of the British Museum*)

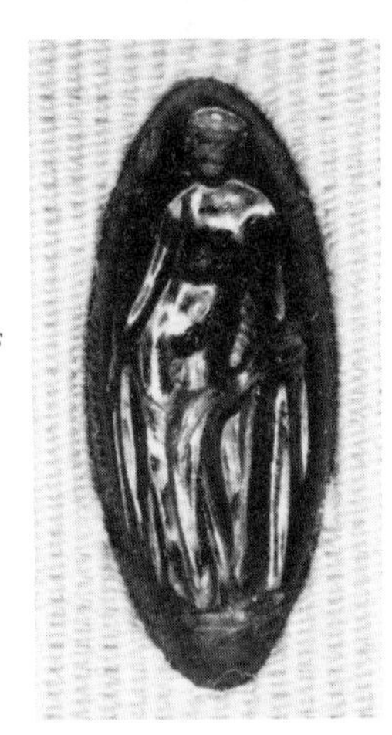

Figure 2 (*right*) Garnet intaglio seal in elongated oval shape (21 × 9 mm). The carving represents a youthful Dionysos with drapery wrapped around his legs. He leans against a column and holds a *thyrsos* in his right hand. The garnet is cabochon-domed rather than flat-cut and the date Hellenistic. BM gem collection #1162. (*Photo by author; courtesy of the Trustees of the British Museum*)

Figure 3 (*above, right*) Garnet intaglio seal stone rendered in a cabochon-domed cut with a very high polish. The figure is a young Satyr holding a shepherd's crook on left shoulder. His right hand is wrapped in a panther skin. The bottom of this stone was cut concave and highly polished, no doubt to enable the light to bring out the colour of the garnet more effectively. The date of the stone is Hellenistic and measures 17 × 13 mm. BM gem collection #1163. (*Photo by author; courtesy of the Trustees of the British Museum*)

Figure 4 (*left*) Garnet intaglio seal mounted in a Graeco-Roman gold ring. This stone reveals a popular myth, particularly in Corinth, of Bellerophon charging with a spear on Pegasus, the flying horse. The garnet is not large, measuring 9 × 11 mm. BM gem collection #1914. (*Photo by author; courtesy of the Trustees of the British Museum*)

Figure 5 (*right*) Rare grossular garnet cameo mounted in a Graeco-Roman gold ring with open-back setting. It represents the head of a boy (Ganymedes) wearing a Phrygian cap with long flaps. The stone is flawless, very transparent and brownish-orange in colour. The open back mounting reveals that the stone is concave cut and highly polished. The date is undetermined, but probably Hellenistic. The stone measures 12 × 15 mm in the unusual triangular shape. BM gem collection #3424. (*Photo by author; courtesy of the Trustees of the British Museum*)

Figure 6 (*left*) Hellenistic earring that appears to have been popular over several centuries, from about the third century BC to Imperial Roman times. The use of the garnet here is simply decorative in function. The stone is domed rather than flat-cut and well polished. In the pear-shaped bezel above the crescent-shaped garnet the missing stone was also probably garnet. The goldsmithing talent reveals a high degree of craftsmanship. The use of granulation and fine wire filigree as well as the intricate wire bangles are techniques which are typical of the period. The tiny amphora is made in stamped sheet gold which was made in two halves and soldered together. The fine wire filigree and the tiny granules decorating the amphora are particularly intricate. Flanking the pear shaped bezel two dolphins are seen, created in very delicate filigree and granulation. BM jewellery collection #2357. (*Photo by author; courtesy of the Trustees of the British Museum*)

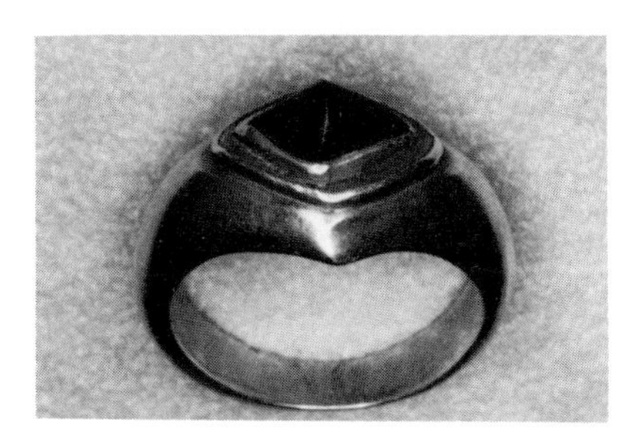

Figure 7 Lozenge-shaped garnet mounted in solid gold cast ring of Graeco-Roman date. The garnet is purely decorative in this use. This ring was probably a man's ring, for the ring size (size 12) and massive style suggests this function. BM finger ring collection #755. (*Photo by author; courtesy of the Trustees of the British Museum*)

Figure 8 Gold wire finger ring with three garnets. The wire is constructed to appear like serpents, whose heads emerge on each side of the two pear shaped garnets. The ring is designed to be worn lengthwise on the finger. BM finger ring collection #771. (*Photo by author; courtesy of the Trustees of the British Museum*)

Nevertheless, the gods continue to be seen on intaglio seals throughout the period. A youthful Dionysos can be seen on one garnet seal fashioned into an elongated oval (21 × 9 mm) in *Figure 2*. The artistic craftsmanship is less than fine, but the detail is quite good.

Hellenistic gem cutters also learned to undercut their stones so that the colour might be more visible in the settings. In one intaglio of a young Satyr (*Figure 3*) the bottom of the stone is cut concave. In some samples of garnet seals in finger rings, a very fine red colour can be seen as the light travels through the stone and reflects off the gold backing. Most of the garnet seals in the British Museum do have polished backs, although not many display a concave bottom. However, it is not always possible to discover this feature since many finger rings and other jewellery items incorporate closed-back settings. The young Satyr garnet is highly polished on both sides of the stone.

Favourite local myths also appear on garnet intaglios. In *Figure 4* Bellerophon can be seen charging with spear, riding Pegasos. This myth was particularly popular in Corinth. The garnet (9 × 11 mm) is set in a gold Graeco-Roman ring of simple design.

The use of garnets for intaglios in the Hellenistic period was prolific. The British Museum has many dozens of garnet samples in the gem collection, and many more are found in the finger ring collection. However, it is rare to find garnets cut in cameos. Such a stone was discovered in the British Museum gem collection (*Figure 5*). It displayed the head of a boy (perhaps Ganymedes) wearing a Phrygian cap with long flaps. The garnet was unusual in that it was not the usual almandine, but it was a grossular garnet (hessonite), quite flawless and transparent, and brownish-orange in colour. Also, it was highly polished and the bottom of the stone was concave cut (visible in the open setting).

The use of garnet gems as decorative items in jewellery also became popular in Hellenistic times. Garnets and other gemstones were used as animal eyes, collars and diadems in earrings particularly. Occasionally, garnets would be cut into amphora pendants, beads, balls and variously shaped cabochons. *Figure 6* presents a very popular class of earring throughout the Graeco-Roman period, but especially in the Hellenistic period. The garnet is crescent-shaped, repeating the same shape in the gold and filigree of the earring. The garnet is quite decorative in this example, perhaps adding a dimension of colour to the ensemble.

The decorative use of garnets accelerates in the Roman period. A garnet finger ring, possibly of Roman date exemplifies this trend (*Figure 7*). The stone is cut into a lozenge-shaped cabochon and mounted into a heavy gold cast finger ring. The stone is very dark and transmits no colour from the backing. Another finger ring of late Hellenistic or early Roman date is composed of three garnets in a mounting of gold wire and filigree (*Figure 8*). The workmanship appears to be somewhat crude when compared to similar jewellery of the period. But the era of the Romans is known for experimentation, and unusually awkward jewellery types are commonly found in the period.

As the Roman period developed, the gold for jewellery became more scarce and precious. Coins of gold replaced gemstones in finger rings, and the coin rings and other gold jewellery were used as barter and often hoarded. Garnets continued to be used throughout the Roman period, particularly as decorative stones in medallions. The garnet medallions of the late Roman period still used cabochon cut stones, but a new tradition seemed to develop from south Russia: the flat cut garnet mounted in cloisonne jewellery. At the beginning, the flat cut stones were somewhat large, glued in place, but polished on both top and bottom.

During the Dark Age period, the kings retained the goldsmiths who had the ancient traditions passed down to them from one generation to another. The flat-cut garnet cloisonne jewellery reached an extremely high plateau of both intricacy of design and quality of workmanship. The finest garnet cloisonne jewellery ever made came from the workshop of the Sutton Hoo master.

The jewellery was discovered in 1939 when a ship burial of the early 7th century was excavated in East Anglia. There were approximately forty-five pieces of cloisonne jewellery in the burial. The jewellery collection incorporated approximately 4000 garnets which were individually cut and polished to fit exactly into the gold cloisons. No glue was used to hold the garnets; they were suspended at the top of each cell, held in place only by the tightness of the fit. The technique also exhibited a high degree of knowledge about light reflecting through the garnets, for each garnet stone was cut very thin and gold backing produced an exceptionally fine red colour. The smallness of some stones in the collection was also surprising. Some were under 1 mm in diameter, and intricately shaped. No other jewellery was to carry on this tradition in later historical periods. It seems it all died with the owner of the jewellery in 622 AD.

Index